Editorial Policy
for the publication of monographs

In what follows all references to monographs, are applicable also to multiauthorship volumes such as seminar notes.

§1. Lecture Notes aim to report new developments - quickly, informally, and at a high level. Monograph manuscripts should be reasonably self-contained and rounded off. Thus they may, and often will, present not only results of the author but also related work by other people. Furthermore, the manuscripts should provide sufficient motivation, examples and applications. This clearly distinguishes Lecture Notes manuscripts from journal articles which normally are very concise. Articles intended for a journal but too long to be accepted by most journals, usually do not have this "lecture notes" character. For similar reasons it is unusual for Ph. D. theses to be accepted for the Lecture Notes series.

§2. Manuscripts or plans for Lecture Notes volumes should be submitted (preferably in duplicate) either to the editor of the series or to Springer-Verlag, Heidelberg. These proposals are then refereed. A final decision concerning publication can only be made on the basis of the complete manuscript, but a preliminary decision can often be based on partial information: a fairly detailed outline describing the planned contents of each chapter, and an indication of the estimated length, a bibliography, and one or two sample chapters - or a first draft of the manuscript. The editor will try to make the preliminary decision as definite as he can on the basis of the available information.

§3. Final manuscripts should be in English. They should contain at least 100 pages of scientific text and should include
- a table of contents;
- an informative introduction, perhaps with some historical remarks: it should be accessible to a reader not particularly familiar with the topic treated;
- a subject index: as a rule this is genuinely helpful for the reader.

Further remarks and relevant addresses at the back of this book.

Lecture Notes in Biomathematics

90

Simon N. Wood   Roger M. Nisbet

Estimation of Mortality Rates in Stage-Structured Population

Springer-Verlag

Berlin Heidelberg New York
London Paris Tokyo
Hong Kong Barcelona
Budapest

Authors

Simon N. Wood
Centre for Population Biology
Ascot Berks, SL5 7QY, U.K.

Roger M. Nisbet
University of Strathclyde
Department of Statistics and Modelling Science
Glasgow, G1 1XH, U.K.

Mathematics Subject Classification (1980): 92D40, 92D25

ISBN 978-3-540-53979-7   ISBN 978-3-642-49979-1 (eBook)
DOI 10.1007/978-3-642-49979-1

© Springer-Verlag Berlin Heidelberg 1991
2146/3140-543210 - Printed on acid-free paper

PREFACE

The stated aims of the Lecture Notes in Biomathematics allow for work that is "unfinished or tentative". This volume is offered in that spirit. The problem addressed is one of the classics of statistical ecology, the estimation of mortality rates from stage–frequency data, but in tackling it we found ourselves making use of ideas and techniques very different from those we expected to use, and in which we had no previous experience. Specifically we drifted towards consideration of some rather specific curve and surface fitting and smoothing techniques. We think we have made some progress (otherwise why publish?), but are acutely aware of the conceptual and statistical clumsiness of parts of the work. Readers with sufficient expertise to be offended should regard the monograph as a challenge to do better.

The central theme in this book is a somewhat complex algorithm for mortality estimation (detailed at the end of Chapter 4). Because of its complexity, the job of implementing the method is intimidating. Any reader interested in using the methods may obtain copies of our code as follows:

Intelligible Structured Code

1. Hutchinson and deHoog's algorithm for fitting smoothing splines by cross validation
2. Cubic covariant area–approximating splines
3. Cubic interpolating splines
4. Cubic area matching splines
5. Hyman's algorithm for monotonic interpolation based on cubic splines.

Prototype User–Hostile Code

6. Positive constrained interpolation
7. Positive constrained area matching
8. The "full method" from chapter 4
9. The "simpler" method from chapter 4.

In addition we can supply executable copies of the 3–D plotter used to produce the figures. This will screen dump to an Epson–FX compatible printer.

Items 1–9 above are available only as Turbo Pascal (version 5.0 or 5.5) source code. The problems in translating to other dialects of Pascal are minor except for items 8 and 9. The code is supplied free and without any guarantees(!). However we request payment of US \$20 or 10 sterling to cover disks, postage and hassle.   Requests for code should be addressed to R.M. Nisbet, Department of Statistics and Modelling Science, University of Strathclyde, Glasgow G1 1XH, Scotland.

Many people contributed time and ideas to this project. Bill Gurney and Steve Blythe collaborated in previous work based on stage structure models; much of their insight and creativity carries over to the present work. Ann Jones, Andre de Roos and Hans Metz contributed ideas on instability and ill–conditioning that significantly influenced chapter 2. We thank Peter Maas for computational advice and for a beer mat rediscovery of quadratic histosplines! We thank John Gamble and Steve Hay (who is co–author of chapter 6) for access to unpublished data on Loch Ewe copepods, and for remaining supportive during a very long period over which we made little visible progress. Comments by Hal Caswell, Saran Twombly and Nelson Hairston at a Conference on "Future Directions in Zooplankton Biology" in Michigan (October 1989) helped place the work in a broader perspective. We thank Jerome Casas, Steve Hay, Joe Horwood, Hans Metz and Charles Taylor for valuable comments at different stages of the manuscript preparation. We particularly appreciated a thorough reading by Hans Metz who checked much of the algebra, though taking our lead from modern terrorists we claim responsibility for the remaining errors. Kay Johnston battled with illegible handwriting and with text from two different word processors used on several, subtly different computers to produce the final manuscript.

Most of the work was carried out at the University of Strathclyde and we thank our colleagues in the Department of Physics and Applied Physics and in the Department of Statistics and Modelling Science for their interest. The work was completed while SW was employed first at the MAFF Fisheries Laboratory, Lowestoft, and latterly at the Centre for Population Biology, Imperial College. Financial support came from the UK Natural Environment Research Council through grant GR 5210 and a studentship held by SW.

Simon Wood   University of Strathclyde
Roger Nisbet   October 1990

CONTENTS

CHAPTER 1

INTRODUCTION

1.1 Inverse problems in population ecology

A major aim of population ecology is to explain observed patterns of plant and animal abundance. At the single species level, one approach is to construct mathematical models of abundance whose parameters are estimated in a manner quasi–independent of the data which it is hoped to explain. Disagreement between such a model and observed population data implies that the model is false, agreement that it may be correct. Mathematically, this approach to explaining abundance yields a 'direct' problem: the model and its parameter estimates constitute a mathematically complete system which can be solved.

The direct approach has been used successfully by many workers to interpret data on laboratory populations, and has potential as a tool for linking laboratory and field data (e.g. Nisbet *et al.*, 1989). However there are many difficulties, of which those associated with mortality estimation are possibly the most fundamental. Plants and animals die from a worryingly diverse set of causes; if we have to try and explain a set of population data by hypothesising too many (poorly understood) causes of mortality, then there is a good chance of obtaining some very trendy dynamics without very much insight.

One way of by–passing the difficulties of the direct approach is to try and solve an 'inverse problem'. Given some fairly general modelling assumptions can we extract patterns of recruitment and mortality most likely to explain a given set of population data? Unlike the direct problem, such problems are usually underconstrained.

Ecological inverse problems answer different questions to those requiring the direct approach. A mortality pattern derived by solving an inverse problem may "explain" a set of population data in the sense of enabling a reconstruction of the original data which is not too offensive to a friendly statistician, but on its own it says nothing about the physiological and environmental factors which are responsible for this mortality. However there are many situations (e.g. production calculations) in which ecologists want a quantitative characterisation of mortality rates which is independent of any assumption about the root causes. Such information may of course be used subsequently to identify possible causes, for example by relating zooplankton death rates to growth rates of larval fish.

Some inverse problems are hopelessly underdetermined. For example, any population trajectory can be produced by an infinite family of time–dependent *per capita* birth and death rates. Mathematically this can be seen by considering a population, $\eta(t)$, described by

$$\frac{d\eta(t)}{dt} = [b(t)-\mu(t)]\eta(t) \qquad \text{with } \eta(0)=\eta_0 . \qquad (1.1.1)$$

Given data points sufficiently close to permit a reasonable estimate of $d\eta/dt$, it is straightforward to estimate the *difference* $b-\mu$, but impossible to separate these quantities.

Much more interesting is the situation where the inverse problem is still underdetermined, but where with additional assumptions it is possible to extract information on mortality rates. For example, suppose an experimentalist can identify individuals in the above population as being either juvenile or adult with the two subpopulations being denoted by $\eta_1(t)$ and $\eta_2(t)$: further suppose that individuals recruited to the juvenile sub-population η_1 at time t have a *known* rate (m) of maturation to the adult sub-population. (The knowledge of this rate might come for example from laboratory experiments.) Then if

$$\frac{d\eta_1(t)}{dt} = b(t)\eta_2(t)-[\mu_1(t)-m]\eta_1(t) \qquad (1.1.2)$$

$$\frac{d\eta_2(t)}{dt} = m\eta_1(t)-\mu_2(t)\eta_2(t) \qquad (1.1.3)$$

the problem of death rate estimation is still underdetermined since at each time, there are three quantities (b, μ_1, μ_2) to be calculated from two equations. However if there is a known relationship (e.g. equality) between the juvenile and adult death rates, these can now be determined.

The solution of the above inverse problem had two components: *new information* (the value of the parameter m), and a *new assumption* (equality of *per capita* death rates for juveniles and adults). The methods to be developed in this monograph recognise that effective use of both is the key to progress with mortality estimation problems.

1.2 Copepod mortality rate estimation

One area of ecology in which mortality estimation techniques have been, and continue to be, in great demand is the study of zooplankton populations (McCauley *et al.* 1990), in particular marine copepods whose development involves a sequence of about 12 morphologically well defined stages known as nauplii (usually stages 1-6) and copepodites (usually stages 7-12). Copepods are among the principal consumers of oceanic primary production, and are the food for larvae of many species of commercially important fish. Despite extensive research in many areas of copepod biology, understanding of the relationship between copepods and their predators remains largely qualitative. One reason for this is that it is not possible to distinguish between individual naupliar or copepodite stages when analysing fish stomach contents. Thus information on predation rates must come in part from the assessment of mortality rates in naturally occuring copepod populations.

One example of the need for good predation estimates is the testing of Hjort's (1914) 'Critical period' hypothesis. This asserts that the strength of a year class of fish is determined by the survivorship of larvae during the 'critical' few days after the exhaustion of yolk reserves. This survivorship is assumed to depend on the availability of suitable food, i.e. copepods. Any attempt to validate this hypothesis will require time- and age-dependent mortality rate information for the supposed food species in the presence of fish larvae.

Ecotoxicology is another area in which mortality estimates are important: at what time of year and at what age are populations most susceptible to pollutant stress? Is a reduced population in the field the result of reduced natality or increased mortality? These are questions which may not easily be addressed by extrapolation from laboratory studies, since they are likely to be affected as much by other elements of the ecosystem as by the physiology of the animals in question.

Having established the importance of knowing something about copepod mortality rates, it is important to appreciate the problems of data collection from patchily distributed indistinct populations of minute mobile organisms in the open oceans. Because of these problems, biologists wanting to probe the details of copepod ecology have often performed experiments in large marine enclosures (e.g. Grice and Reeve 1982), an approach which reduces the problems caused by mobility and large scale patchiness, but not those associated with small scale aggregation. It is data from enclosed populations that motivated this study. However, even in an enclosed population corpses are impossible to count, thus precluding direct measurement of death rate. Dead copepods may have been eaten or may otherwise fall from the water column. Direct measurement of birth rate is almost as difficult in the absence of egg mortality data and fertility rates.

Practical attempts to estimate mortality must take an indirect approach. Copepods develop through a series of clearly identifiable life stages, the durations of which are believed to be roughly fixed. It is feasible to sample the stage populations to obtain time series of numbers in each stage. A mathematical procedure is then devised to estimate mortality rates by comparison of the time series.

Indirect mortality estimation has three main difficulties:

1) Finite populations and limited resources for sample analysis combine to ensure that the data is sparse and noisy and often unaccompanied by any estimate of expected sampling error.

2) Older stages are more mobile than younger stages so that sampling efficiency is likely to be different for the different stages: older stages may be more able to avoid sampling gear.

3) Even given perfect sampling, estimation of mortality from stage structured population data is non-trivial – see chapter 2.

1.3 The theoretical problems of mortality estimation

If it is possible to assume that within each stage all individuals suffer the same mortality rate then, in order to estimate mortality from stage population estimates, stage durations for all pre–adult stages and recruitment estimates for at least one stage must be available. These requirements are the minimum for the case of data sampled without error, but as shown in chapter 2, they may still be inadequate if the data are noisy. Unfortunately for the experimentalist these minimum requirements are usually beyond reach because of the difficulty of obtaining recruitment estimates. It is thus necessary to construct models of the population under study which introduce in analysis constraints absent in the original data.

It is the search for reasonable assumptions and ways of implementing them in an estimation scheme which has exercised most workers on mortality estimation, but the equally important fact that most of the data are very noisy appears to have received little consideration. The difficulties here are twofold. Firstly, error estimates are rarely produced. Secondly, the variability in spatial distribution of the sampled animals means that even the shape of the error distribution is difficult to ascertain. For example, the distribution of sample sizes expected from a population aggregated in a few large patches will be dominated by the distribution of probabilities of sampling a given number of these patches. This will produce a very different distribution to that expected from a population spread uniformly through the water column.

Even if an apparently satisfactory model is constructed and sampling errors are of known distribution and small magnitude, there are still the difficulties highlighted in section 1.1: Asknes and Magnesen (1988) point out that since a change in stage population caused by increased recruitment is all but indistinguishable from the change occasioned by decreased mortality within the stage: 'a close fit between model calculations and observations does not imply that the parameters are accurately estimated'.

1.4 Existing methods for mortality estimation

Egg ratios

The earliest methods of mortality estimation to receive widespread attention in zooplankton work were the methods based on the 'egg ratio' technique of Edmondson (1960, 1968). In methods of this type single parameters are sought to represent birth and death rate in a population assumed to be undergoing exponential growth (or decay), with steady age structure. The extra information characterising all mortality estimation techniques is in this case provided by recruitment estimates based on the 'egg ratio': the ratio of eggs to adults, which must be determined experimentally. An estimate of the hatching rate of these eggs is also required. The techniques consist of formulae relating the measured egg ratio, population and hatching rate to the instantaneous birth rate and hence to the instantaneous mortality rate.

The drawbacks of this approach lie in the difficulty of measuring the egg ratio, or indeed any other measure of recruitment, coupled with the restrictive modelling assumptions of exponential growth and steady equilibrium age structure. Seitz (1979) addressed the latter critisisms by developing a generalisation of Edmondson's method based on a multi–stage model. Although his method shared the assumption of equilibrium age structure with previous methods, Seitz tested the technique against simulated data in which death rate was allowed to vary to produce population cycles. Whilst reporting mixed success in reconstructing birth rates from this data, he does not report attempts to reconstruct death rate.

Paloheimo has addressed a number of problems concerning the details of these methods (Paloheimo, 1974) and has applied them to the study of *Daphnia pulex* (Paloheimo and Taylor, 1987), a species for which they are particularly well suited, since *Daphnia* eggs are carried in a brood pouch by the adult female. This work has not considered the dynamic variation in death rate, or the reliability of the estimates produced in the likely event that the model assumptions are violated.

Cohort analysis

Rigler and Cooley (1974) further developed and corrected a method proposed by Comita (1972) which estimates both stage duration and between–stage survival from datasets displaying clear cohorts. Their method hinges on estimating the mean time of occurrence of a cohort within a stage (the 'temporal mean'). The differences between the temporal means in adjacent stages can be used to estimate stage durations provided one stage duration is already known. Survival in a stage is estimated from the ratio of throughput estimates for the stage itself and the previous stage. Throughput is asserted to be the area under the population/time curve divided by the estimated stage duration. Although with careful application this method produced convincing results when applied to Rigler and Cooley's experimental data, they failed to test its efficacy on simulated data. This deficiency was later rectified by Saunders and Lewis (1987) who showed by numerical simulation that the method is effective if through–stage survivorship is higher than 0.5 and stage durations increase from stage to stage by not more than 50%.

Notwithstanding considerable popularity (e.g Burns 1980, Hairston *et. al.* 1983, Twombly 1983) the method was criticised by Hairston and Twombly (1985). They argued that failure to take into account mortality in the calculation of stage durations severely compromises the method's usefulness, but their proposed solution requires prior knowledge of the mortality rate, rendering it next to useless. Asknes and Hoisoeter (1987), building on the suggestion of Manly (1977), proposed an alternative to Hairston and Twombly's method. This dealt with the latter's objections to Rigler and Cooley's approach, but produced mortality estimates rather than requiring them in advance. The modified technique was used by Asknes and Magnesen (1988) to investigate the population dynamics of four species of copepod in

Lindaspollene, Norway. Once again the study does not assess method accuracy by simulation.

Cohort analysis remains a simple and powerful method of analysis when a population develops in distinct cohorts and data are relatively free from error. Unfortunately this situation of convenient dynamics and ideal sampling is extremely rare. In addition it should be noted that strongly time–dependent mortality would produce shifts in the mean position of a cohort within a stage, which would in turn bias stage duration estimates.

Systems identification

In an effort to deal with both the problem of noisy data and the uncertainties surrounding the underlying dynamics of most copepod populations Parslow, Sonntag and Matthews (1979) suggested using the technique of systems identification to estimate mortality rates and stage durations. The idea is that a population model is constructed which is then fitted to the data using some non–linear least squares procedure, in order to obtain best fit model parameters. Parslow et. al. fitted four different models to simulated noisy data from a fifth model, hence ascertaining the robustness of each model to violation of its own assumptions. The technique was very successful as a means of estimating production (Sonntag and Parslow, 1981) but worked less well for mortality estimation (Parslow et. al. 1979).

Despite the reservations of its original proponents, systems identification has found favour in several marine mesocosm studies (e.g. Harris et al. 1982; Hay, Evans and Gamble, 1988). For some datasets the method produces a plausible set of best fit stage population trajectories and hence an arguably 'believable' set of parameter values; but as Asknes and Magnesen (1988) point out, the logic of such deductions may be flawed. Furthermore scant consideration appears to have been given to the robustness of these estimates to erroneous modelling assumptions, a point which is considered further in section 5.2.

Linear regression methods

Fransz (1980) and Manly (1987) have both proposed mortality estimation methods which fit models by linear regression rather than by the non–linear model fitting methods employed in systems identification. This leads to the elimination of problems caused by multiple minima and convergence failure.

Fransz's method involves fitting a discretised version of one of the models ('linear transfer') of Parslow et. al. (1979) for each stage separately. Recruitment to the stage is described by a development rate which multiplies the previous stage population. Development rate from the stage and mortality rate within the stage are similarly represented by single multiplicative parameters. Fitted to a pair of stages by linear regression, the model yields estimates of the development parameter for the first stage and the sum of development parameter and the death rate for the second stage. Applying this process to all stages produces a set of parameter estimates from which the stage specific mortality rates are calculated. Fransz used this procedure to estimate mortality rates and development rates in a population of copepods but despite obtaining a number of negative death rate estimates, he did not attempt to test the technique's veracity using simulated data.

Manly's method fitted a similar model to a whole dataset. He showed how to express the total population at one sample time in terms of a linear combination of the unknown recruitment parameters and the product of the stage population at the previous sampling time with the unknown stage specific survival probability. The survival probabilities and recruitment parameters can be estimated by linear regression. Manly provided an example, using springtail data. He tested the reliability of the technique by perturbing the original data with artificially generated noise and recording the new parameter estimates. This proccess was repeated to build up a picture of the distribution of the estimated parameters. He also employed the estimates which he obtained to simulate the population for comparison with the original data.

Caswell and Twombly (1989) proposed a modified linear regression method for problems involving time–dependent mortality rates. They found that the matrices to be inverted were frequently ill–conditioned and proposed an algorithm involving "regularisation", a procedure closely related to smoothing. We discuss this method further in chapter 7.

Modified systems identification

Schneider and Ferris (1986) produced an algorithm intermediate between the work of Fransz (1980) and systems identification. They produced a numerical, stage–structured population model with distributed stage durations and a fecundity term linking recruitment to the adult population. The fecundity term must be estimated independently. The model is fitted to the data using an algorithm which first builds up coarse duration estimates stage by stage and then iteratively refines these estimates whilst searching for the survivorship parameters. No results are presented to show the method working with simulation data.

Time– and stage–dependent mortality estimation

All the techniques discussed so far produce time–independent mortality rate estimates, the sole exception to this being the model used by Hay *et. al.* (1988) which represented stage independent mortality as a linear function of time. Such modelling assumptions are restrictive and it is far from clear how mortality estimates should be interpreted when derived from assumptions such as these which are unlikely to reflect reality.

Hiby and Mullen (1980) attempted to estimate stage–specific, time–dependent, mortality rates by fitting a discrete model based on Leslie matrices to the data. Their method requires prior knowledge of both stage durations and fecundity. It produces a time–dependent estimate of the range of survivorship values consistent with the population data, given that the initial population age distribution is unknown. Hiby and Mullen tested their method using simulated data and reported that in the absence of good fecundity information the method was unreliable, a problem we shall discuss in chapter 2. In the copepod studies performed to date, fecundity information, where available, is unlikely to be good enough to justify use of this method.

Lynch (1983) also published a method purporting to estimate mortality rates from stage–structured population data, given stage duration estimates. The method relies on estimating the population that would be expected in a stage at a given time, in the absence of mortality in the stage, and comparing this to the population actually observed. The expectation value is obtained using the assumption of uniform age distribution in preceding stages to estimate recruitment to the stage. This recruitment figure is reduced by an amount determined by the unknown mortality within the stage. Maturation from the stage is calculated in the same way, thereby producing a non–linear equation which can be solved for the mortality rate. Quite how Lynch derives the equation which he solves is unclear and he does not report testing of the scheme, perhaps on the basis of the bizarre argument that since one must choose the parameters used in simulation the results of tests with simulated data are useless (Lynch, 1983).

1.5   This monograph

In chapter 2 we use a mortality estimator derived by S.P. Blythe from the stage structure models of Gurney, Nisbet and Lawton (1983) to investigate the feasibility of estimating mortality using stage structure models. In response to some fairly ugly instabilities identified in chapter 2, chapter 3 provides a brief introduction to spline functions, which have some useful properties which can help avoid estimator instability. Splines are usually applied to point data, so the second part of chapter 3 derives some apparently new results for fitting cubic splines to aggregate data and dealing with covariance. Chapter 4 uses standard spline theory and the new results to produce a mortality estimation scheme based on the reconstruction of the continuous, time–varying, age structure of a population. In chapter 5 this new method is compared to three of the methods described above, using data produced by numerical simulation. In chapter 6, which is written in collaboration with S. Hay, the new method is applied to some real copepod data from Loch Ewe, Scotland. Chapter 7 summarises our conclusions and outlines possible lines for further work.

CHAPTER 2

MORTALITY ESTIMATION SCHEMES RELATED TO
STAGE STRUCTURED POPULATION MODELS

2.1 Introduction

The formalism developed in recent years for modelling stage structured populations would appear to provide a rigorous and appealing basis from which to approach mortality estimation (van Straalen 1986). This formalism is applicable to models of organisms whose life history is made up of a sequence of clearly identifiable stages, of known but not necessarily equal duration (Gurney, Nisbet and Lawton 1983; Gurney, Blythe and Nisbet 1986). In their simplest form, these models assume that all individuals within a stage have the same vital rates (in particular *per capita* death rate) at a given time, and that stage duration is time−independent. The population of the jth stage, $\eta_j(t)$, at time t changes through recruitment of new individuals to the stage, maturation from the stage to its successor, and death. Mathematically this implies that the stage population obeys the differential equation

$$d\eta(t)/dt = R_j(t) - M_j(t) - \mu_j(t)\eta_j(t) \qquad (2.1.1)$$

where $R_j(t)$ and $M_j(t)$ are respectively recruitment rate to, and maturation rate from, stage j at time t, and $\mu_j(t)$ is *per capita* death rate in the stage. The maturation rate is related to the recruitment rate to, or the age distribution within, the stage at previous times. If we assume that no information at all is available for times prior to t=0, and that the initial age distribution within the stage $f_j(x)$, $0 < x \leqslant \tau_j$, at t=0 is known, then the maturation rate is most conveniently expressed in the form

$$M_j(t) = \begin{cases} f_j(\tau_j - t)P_j^+(t) & 0 \leqslant t < \tau_j \\ \\ R_j(t - \tau_j)P_j(t) & t \geqslant \tau_j \end{cases} \qquad (2.1.2)$$

where

$$P_j(t) = \exp\left[-\int_{t-\tau_j}^{t} \mu_j(x)dx\right], \quad P_j^+(t) = \exp\left[-\int_0^t \mu_j(x)\,dx\right] \qquad (2.1.3)$$

where τ_j is the stage duration. The quantity $P_j(t)$ represents through stage survival probability for an individual recruited at time $t - \tau_j$ and maturing at time t, while $P_j^+(t)$ represents the probability of an individual already in the stage at t=0, surviving to mature at time t ($t < \tau_j$). A population containing n stages will be described by n equations of type (2.1.1), which are coupled by the relation

$$R_{j+1}(t) = M_j(t), \qquad j = 1..(n-1). \qquad (2.1.4)$$

The model described by (2.1.1) to (2.1.4) makes possible in principle the derivation of estimators for continuously varying stage-specific death rate, provided that we have approximations to the stage populations $\eta_j(t)$, the stage durations τ_i, and the recruitment to just one stage ($R_1(t)$ for example). Our aim in this chapter is to investigate whether such an estimator would avoid the pitfalls highlighted in chapter 1. In section 2.2, we look at the simplest possible situation – a single stage from which there is no maturation (mathematically equivalent to an unstructured population)– in order to establish a few general points relating to sampling, interpolation and smoothing. In section 2.3 we present an estimator which preforms acceptably on a single stage to which the rate of recruitment is known with reasonable accuracy. However it turns out that there are fundamental problems associated with the use of such estimators on a sequence of stages; these are discussed in section 2.4, and their implications assessed in section 2.5.

2.2 Preliminaries: mortality estimation for an unstructured population with known birth rate

In this section we address some elementary aspects of mortality estimation which are *not* specific to stage structured populations. We consider an *unstructured* population (or equivalently the adult stage of a structured population) and discuss the effects of noisy data and of discrete sampling. Throughout the section, we develop our arguments through the device of using an artificial dataset, for which we have *chosen* all the parameters, including those we wish to estimate, and then attempting to 'reconstruct' the mortality rate used in the creation of the artificial data.

To create this 'test population' time series we solve (2.1.1) numerically for a single stage without maturation. Recruitment, R(t) is held constant and the *per capita* death rate $\mu(t)$ is assumed to vary sinusoidally:

$$\mu(t) = A + B\cos(\omega t) \qquad (2.2.1)$$

where A, B and ω are positive constants, with $A > B$.

To derive a crude mortality estimator, assume that the population, $\eta(t)$ is 'sampled' at equally spaced times $t_i = i\Delta t_s$ where Δt_s is the sampling interval. Denote by $\Delta\eta_i$ the change in population over the time interval $[t_i, t_{i+1}]$, so that if ρ_i, D_i represent respectively total recruitment and deaths over the same time interval, then D_i is simply $\rho_i - \Delta\eta_i$. The average *per capita* death rate over the same time interval may be approximated by dividing D_i by the average population over the time interval (and of course by Δt_s). If Δt_s is sufficiently small, we can approximate the integral involved in calculating the average population by the trapzoidal rule which amounts to averaging the instantaneous populations at the end points. Thus we obtain as estimator for the *per capita* death rate over the small time interval

$$\bar{\mu}_i = 2(\rho_i - \Delta\eta_i)/[(\eta_i + \eta_{i+1})\Delta t_s].\qquad(2.2.2)$$

Precisely how to use this estimator depends on the nature of the available information relating to recruitment. If, as for example with egg ratio methods, we have an estimate of the instantaneous recruitment rates, then we can make further use of the trapezoidal approximation and write

$$\rho_i = (R_i + R_{i+1})/(2\Delta t_s).\qquad(2.2.3)$$

Sampling interval and interpolation

The results of applying the above method to data sampled at different intervals Δt_s are shown in figure 2.1 and demonstrate (not unexpectedly) that there is an improvement in the precision of the estimates of *per capita* death rate as the sampling interval is reduced. Frequent sampling of real populations is of course expensive and tedious; it is therefore of interest to know whether a similar improvement in precision could be obtained by working with coarsely sampled data (large Δt_s) and interpolating so as to obtain data with a small enough Δt to make efficient use of (2.2.2). For example, with a population sampled every 7 days we could interpolate between points (e.g. by using cubic splines – see chapter 3) and obtain nominal population values every day or every 0.1 day. Fig. 2.2 contains results of one such test and establishes that considerable improvement in death rate estimation is obtainable with judicious interpolation.

Noisy data and smoothing

Real data is not only expensive and difficult to collect; it is normally noisy, and mindless interpolation is likely to produce biologically meaningless, artifactual fluctuations. In chapter 3 we shall discuss systematic approaches to combined smoothing and interpolation of sparse, noisy data, but to illustrate the importance of smoothing we show in Fig 2.3 the results of applying our estimator (2.2.2) to artificial population data perturbed by random error. The new data was processed with a simple smoothing scheme, similar in spirit to the methods of Tukey (1977) and previously used by Evans (1985) for the analysis of zooplankton populations. This involved a three point running median followed by a three point running mean with successive points weighted (1:2:1).

The take–home message from these simple examples is that *if the recruitment rate is known*, then interpolation and smoothing can substantially improve death rate estimation from data with a coarse sampling interval and even quite large sampling error. We discuss this further in chapter 7.

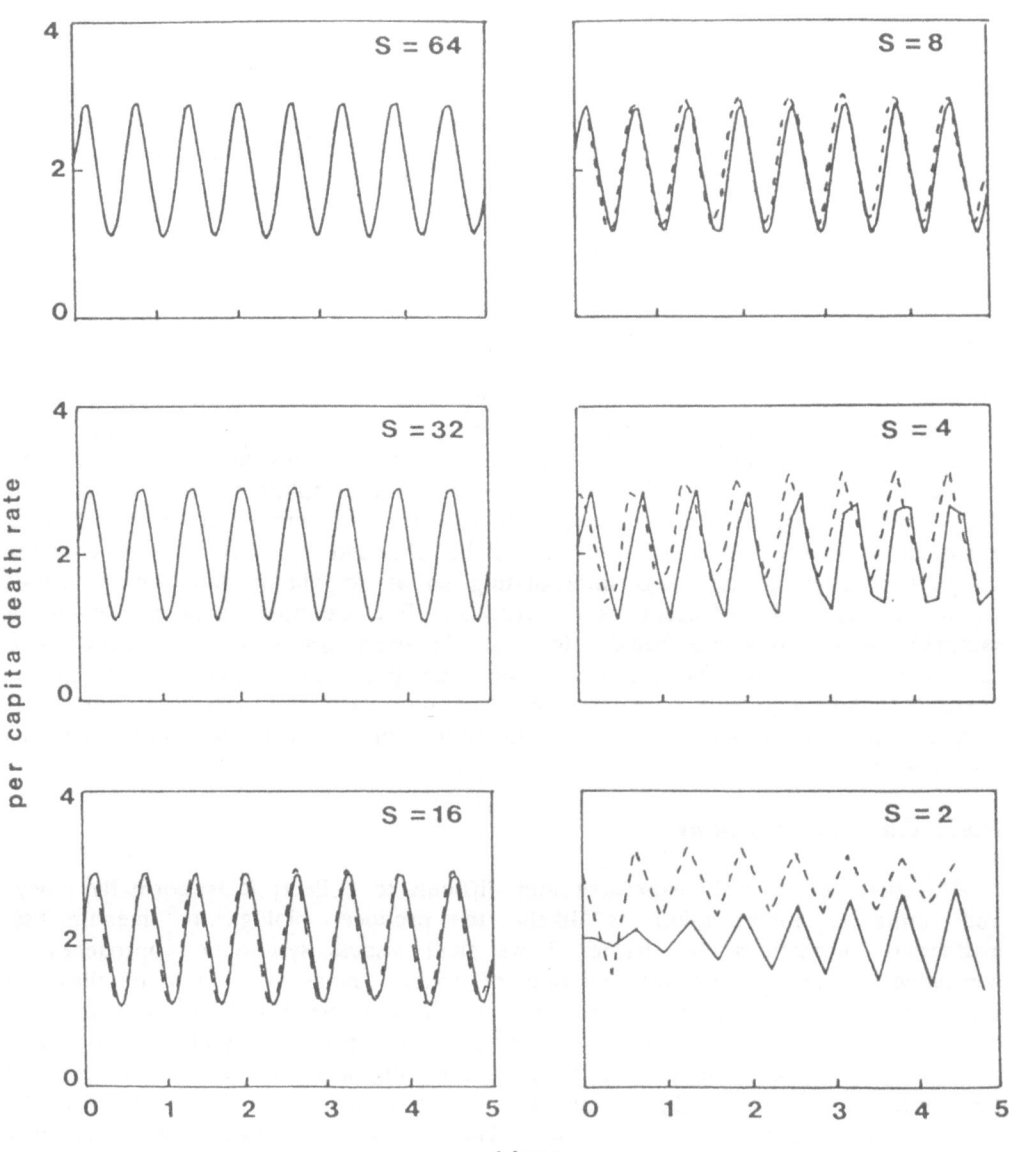

Figure 2.1: Reconstruction of *per capita* death rate from simulated population data for an adult stage experiencing sinusoidal death rate. The population was 'sampled' s times per death rate cycle and the estimator was applied directly to this data. Point estimates of death rates are joined by the broken line, the true death rate values at equivalent times are joined by the continuous lines. The estimation scheme was provided with exact recruitment data.

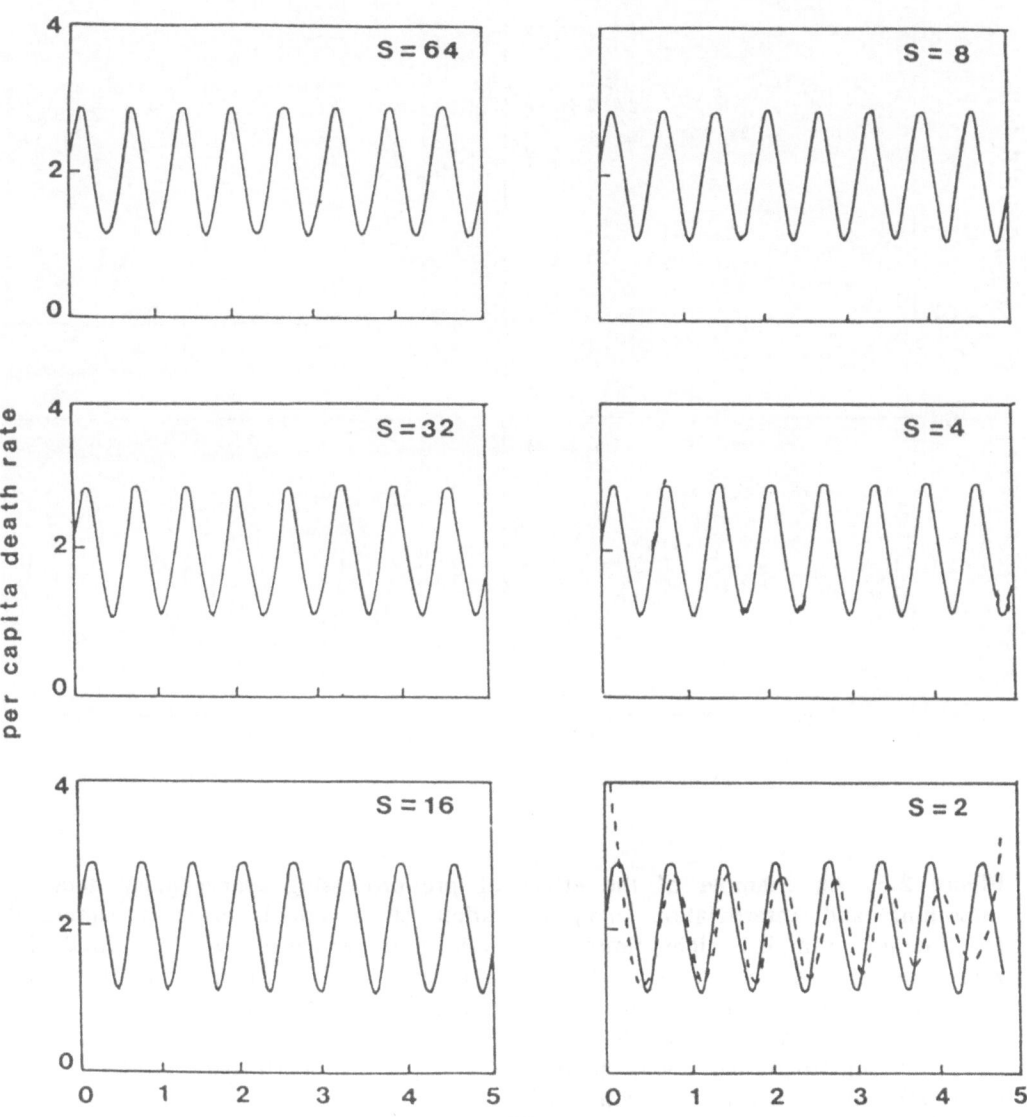

Figure 2.2: The result of applying our estimator to the same data as that used in Figure 2.1, but this time using cubic spline interpolation to reduce the time step used by the estimator. Note the improvement in reconstruction compared to Figure 2.1.

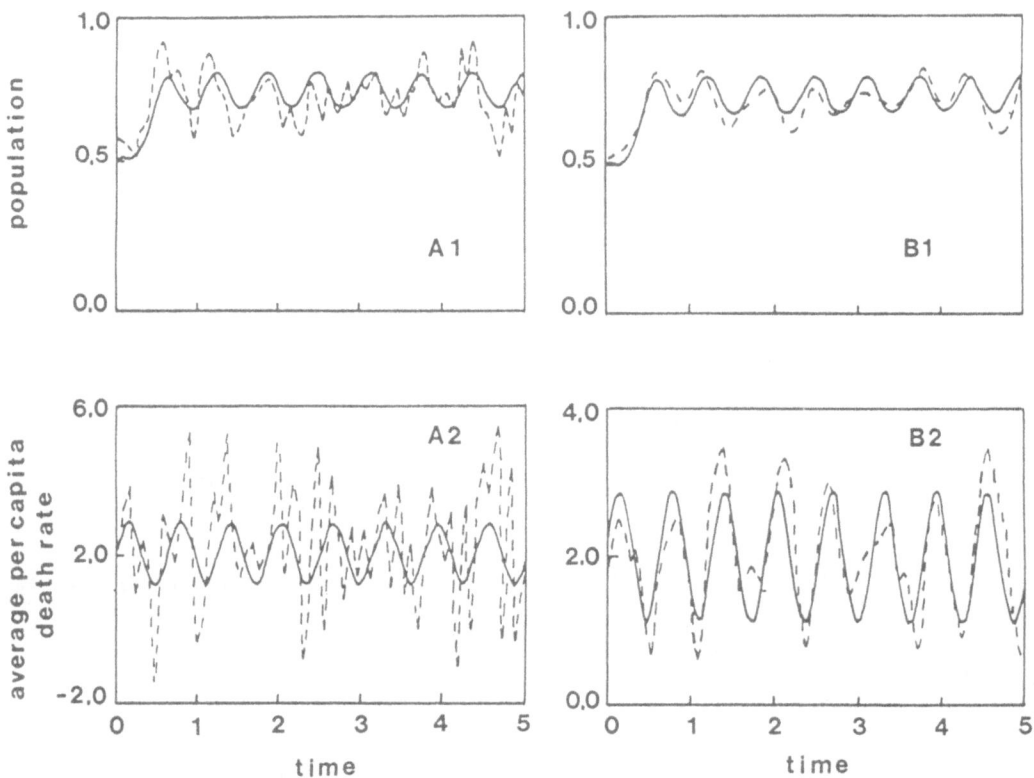

Figure 2.3: An example of the effect of pre-processing sparse noisy data by smoothing and interpolation. A1) An adult stage population was simulated (continuous), sampled eight times per cycle and peturbed by noise (dashed). A2) The sparse, noisy data from (A1) was used without smoothing or interpolation to reconstruct *per capita* death rate. The dashed line is the reconstruction, the solid line the target death rate. B1) The sampled data from (A1) was smoothed and cubic spline interpolation was performed on it (dashed line). B2) Attempted reconstruction of *per capita* death rate (dashed) from the smoothed and interpolated data of (B1) again the target death rate is the continuous line.

2.3 Mortality estimation for a single stage of known duration

Now consider a single stage of duration τ for which the recruitment rate R_i and the stage population η_i are known at a series of equally spaced times given by $t = i\Delta t$ ($i = 0,1,2....m$). Assume that the stage duration is an integer multiple of Δt ($\tau = k\Delta t$). Suppose the initial age distribution f_j ($j = 0, 1 ...$ $k-1$), i.e. the number of individuals whose age (in the stage) is between $j\Delta t$ and $(j+1)\Delta t$, is known. Wood et $al.$ (1989) developed an estimator for the average per $capita$ death rate over the time interval $[i\Delta t, (i+1)\Delta t]$:

$$
\bar{\mu}_i \simeq
\begin{cases}
\dfrac{\hat{R}_i - \Delta\eta_i/\Delta t - \hat{f}_{k-i}P^+_i}{\hat{\eta}_i - \frac{1}{2}\Delta t\, f_{k-i+1}P^+_i} & 0 \leqslant i < k \\[4mm]
\dfrac{\hat{R}_i - \Delta\eta_i/\Delta t - P_i V_i}{\hat{\eta}_i - \frac{1}{2}\Delta t\, R_{i-k+1}P_i/(1-\bar{\mu}_{i-k}\Delta t)} & i \geqslant k
\end{cases}
\tag{2.3.1}
$$

where
$$
V_i = \frac{\hat{R}_{i-k}}{(1-\bar{\mu}_{i-k}\Delta t)}
\tag{2.3.2}
$$

and
$$
P_i = \prod_{j=i-k}^{i-1} (1-\bar{\mu}_j\Delta t), \qquad P^+_i = \prod_{j=0}^{i-1} (1-\bar{\mu}_j\Delta t)
\tag{2.3.3}
$$

The superscript $\hat{\ }$ denotes the average value of a quantity over a time or age interval obtained using a trapezoidal approximation; thus for example

$$
\hat{R}_\ell = \tfrac{1}{2}(R_\ell + R_{\ell+1}), \qquad \hat{f}_{k-\ell} = \tfrac{1}{2}(f_{k-\ell} + f_{k-\ell+1})
\tag{2.3.4}
$$

So, given values of η_i and R_i for all i, and of the initial age distribution within a stage, f_{k-i} for all $1 < i \leqslant k$ it is possible to calculate both μ_i and M_i for all i. This is done by calculating $\bar{\mu}_i$ and P_i (or P^+_i) sequentially for all i from zero upwards. Many similar schemes can be derived − for some discussion see Wood (1989).

The method described by (2.3.1) to (2.3.4) requires that $\mu_i\Delta t \ll 1$ but in experimental studies of copepods, for example, per $capita$ death rates as high as 1 day^{-1} are to be expected in certain stages, whilst even daily sampling is well beyond the scope of most experimental programs. To meet this condition in practice, Δt will have to be much smaller than the sampling interval. This can be achieved by interpolation with cubic splines between data points as already discussed in section 2.2.

2.4. Instabilities associated with mortality estimators

Mortality estimators of the type derived in section 2.3 produce estimates of maturation rates out of a stage as well as of instantaneous death rates. It is here that the apparent power of the approach lies, since maturation from a stage represents recruitment to the next stage. Hence the estimator can in principle be applied sequentially to each stage population in a dataset, the only independent recruitment estimate required being that to the first stage of the sequence (see Fig. 2.4). This approach is highly appealing since the direct measurement of recruitment to every stage is very difficult. Obtaining recruitment information for one stage, however, may be possible from fecundity data or if a very short stage exists which is subject to low mortality. Indeed it is on such assumptions that the 'Egg Ratio' techniques discussed in Chapter 1 are based. In the case of low mortality in a short stage, j, recruitment to the next stage, j+1, could be approximated by the crude formula:

$$R_{j+1}(t) \approx \frac{\eta_j(t)}{\tau_j} \qquad (2.4.1)$$

In the next two subsections we will examine the stability of this sequential approach.

Age propagating instability

Wood *et al.* (1989) considered the propagation through the age structure of a small systematic error, δR_1, in the initial recruitment estimate. For simplicity they considered a population in which all rates and all stage populations are constant and evaluated the error in an estimate of *maturation rate* from a stage induced by an error in the assumed value of the recruitment rate to the stage. We have not managed to extend their proof to the general case where all vital rates vary with time, but the analysis of the special case is still instructive, as it illustrates the fundamental mechanism producing instability in simple schemes invloving a sequence of stages.

Suppose we are supplied with the equilibrium value, η, of a stage population and with the recruitment rate, R, to the stage. Further assume that the stage duration τ is known. The the *per capita* death rate, μ, and the maturation rate, M, are obtained by solving simultaneously the equations

$$R - M - \mu\eta = 0; \qquad (2.4.2)$$

$$M = R \exp(-\mu\tau). \qquad (2.4.3)$$

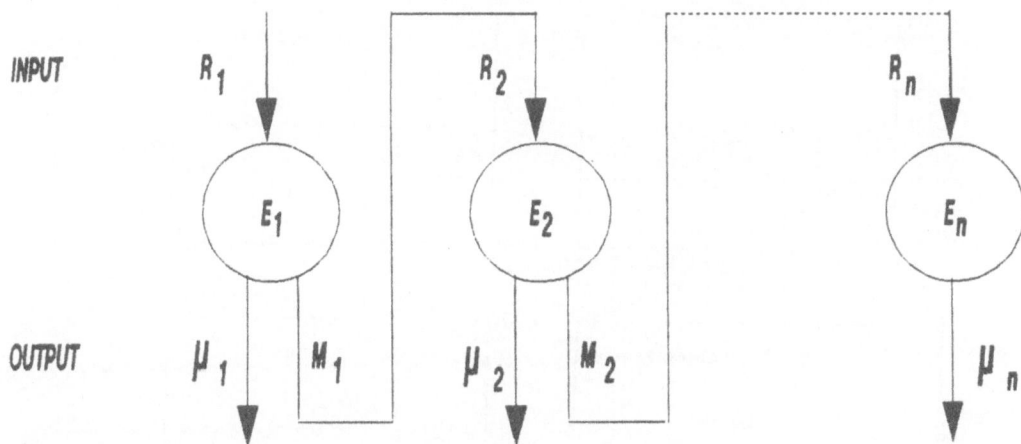

Figure 2.4 Schematic representation of sequential application of a mortality estimator in a stage structured population.

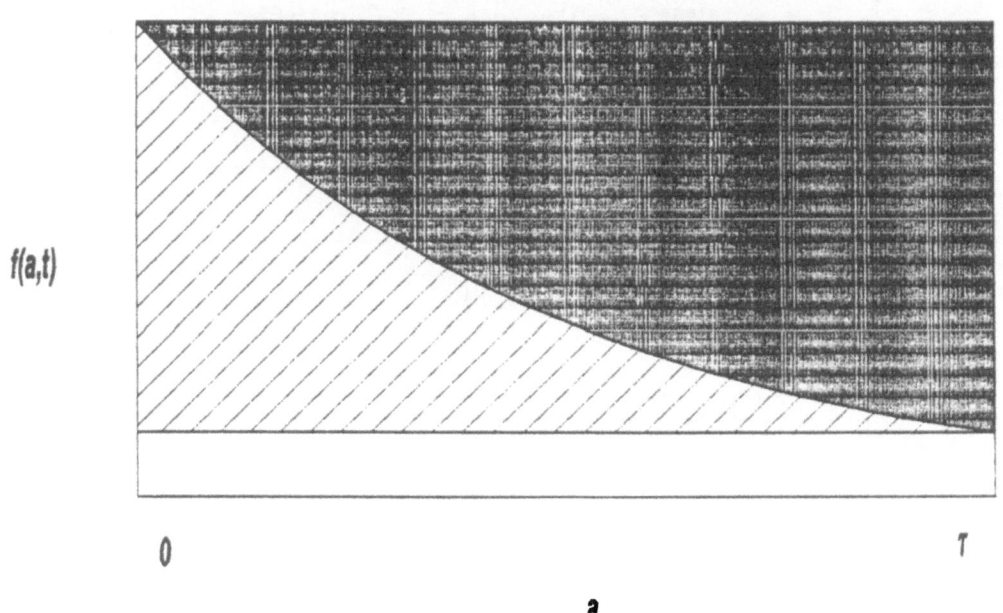

Figure 2.5 The mechanism of the age propagating instability. The age distribution f(a,t), defined formally in chapter 3 has the properties that f(0,t) = R(t), f(τ,t) = M(t), and $\eta(t) = \int_0^\tau f(a,t)da$. Thus the hatched area represents $\eta - M\tau$ and the shaded area represents $R\tau - \eta$. It is obvious that the magnitude of the right hand side of equation (2.4.5) must exceed one.

18

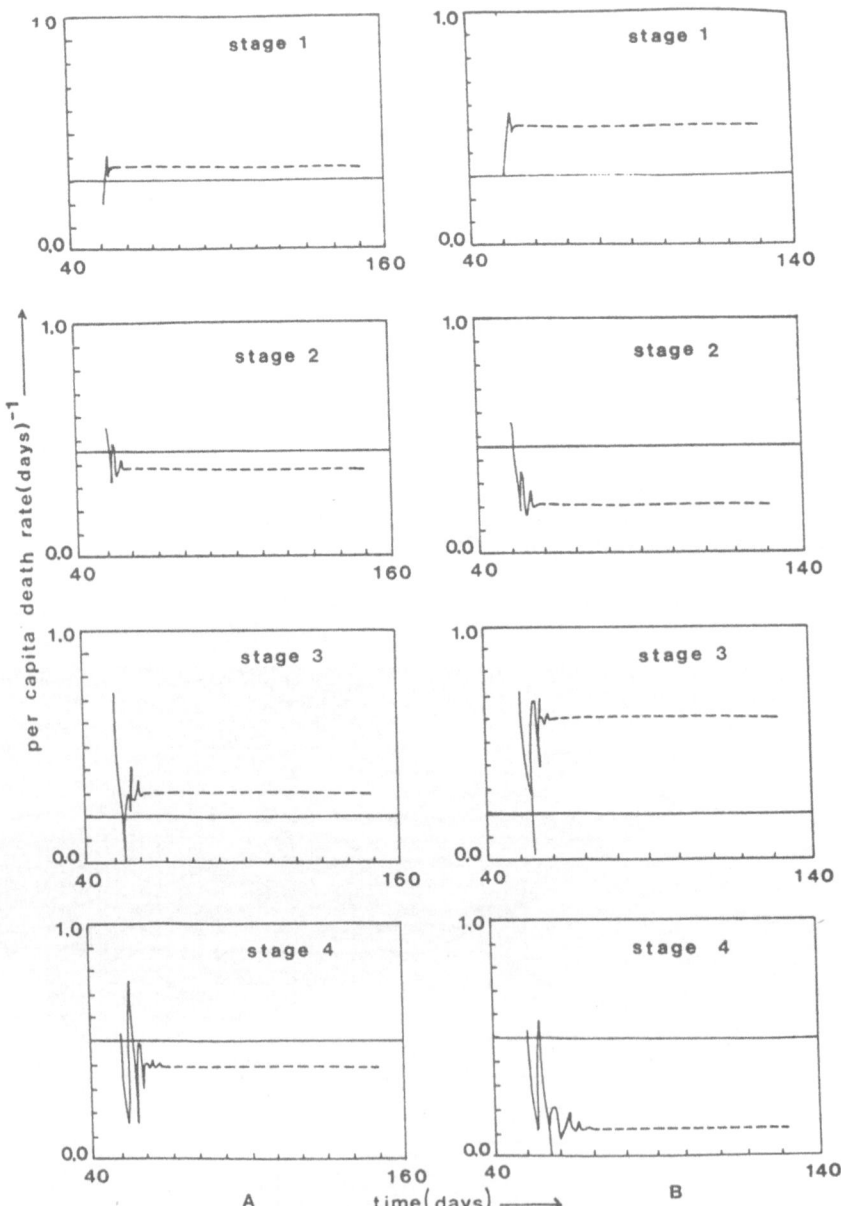

Figure 2.6 Age–propagating instability. A simulated population dataset for four stages with constant vital rates was constructed. The estimator in section 2.3 was applied to the artificial dataset with the aim of reconstructing the original death rates. However the assumed recruitment rate was in error by (A) 10% and (B) 30%. The continuous horizontal lines represent to "correct" death rate, the broken lines, the results of the reconstruction. The figures show the divergence of death rate estimator from one stage to the next. The initial oscillations are a transient arising from the need to assume some initial age distribution with each stage.

Eliminating μ between these equations yields a relationship between the *input* value of R and the *inferred* value of M:

$$R_T = M_T + \eta \{\ln R - \ln M\}. \tag{2.4.4}$$

If we now consider a small error δR in the input value of R, then this produces an error δM in the estimate of M which can be calculated to first order by differentiating equation (2.4.3) and reorganising terms to obtain

$$\frac{\left[\frac{\delta M}{M}\right]}{\left[\frac{\delta R}{R}\right]} = \frac{R}{M}\frac{dM}{dR} = -\frac{R_T - \eta}{\eta - M_T} \tag{2.4.5}$$

It is possible with a little algebra (see Wood *et al* 1989) to show that this ratio is always less than -1; the same result can be established by a graphical argument (see Fig 2.5). Since the rate of maturation from one stage is equal to the rate of recruitment to its successor, the implication is that *recruitment and maturation estimates get worse as the method advances through the age structure.* We term this effect *age propagating instability* since it propagates in the direction of increasing age. Wood *et al.* (1989: see Fig. 2.6) give a spectacular example of rapid divergence of death rate estimates caused by this instability.

Although our derivation of the age propagating instability is based on consideration of a population in which the vital parameters are time–independent, there is no reason for supposing that the problem is going to be any less severe in the time–dependent case. Indeed our own experience using both simulated and experimental time series seems to indicate that time dependence makes things considerably worse.

Time propagating instability

A corollary to the result on age–propagating instability is that if we run backwards through the stages, using a final maturation estimate as our starting point, then errors induced by errors in that final maturation estimate should reduce in magnitude as we move back to the younger stages. Wood *et al.* (1989) therefore looked at a class of estimation methods that takes the stage population, η, maturation estimate, M, (plus age distribution at one time, and stage duration) and produces an estimate of μ and R. Any such method must run backwards in time, but Wood *et al.* showed that when this was done any small error in per capita death rate at one time was amplified so rapidly as to make effective computation useless. They called this problem *time propagating instability.* Their proof of the time propagating instability runs as follows:

Rewriting (2.1.1) we obtain the general mortality estimator :

$$\mu(t) = \frac{R(t) - M(t) - \frac{d\eta(t)}{dt}}{\eta(t)} \tag{2.4.6}$$

To work *backwards* through the age structure R(t) must be written in terms of M(t+τ) :

$$\mu(t) = \frac{\dfrac{M(t+\tau)}{P(t+\tau)} - M(t) - \dfrac{d\eta(t)}{dt}}{\eta(t)} \tag{2.4.7}$$

$$[P(t+\tau)]^{-1} = \exp\left\{ \int_t^{t+\tau} \mu(x) \ dx \right\} \tag{2.4.8}$$

Now consider a function $\epsilon(t)$, the error perturbing $\mu(t)$, so that (2.4.8) becomes:

$$[P^*(t+\tau)]^{-1} = \exp\left\{ \int_t^{t+\tau} \mu(x) \ dx \right\} \exp\left\{ \int_t^{t+\tau} \epsilon(x) \ dx \right\} \tag{2.4.9}$$

where P^* is the function P plus the perturbation caused by ϵ. Hence using (2.4.8) and (2.4.9) substituted into (2.4.7) we obtain:

$$\epsilon(t) = \exp\left\{ \int_t^{t+\tau} \mu(x) \ dx \right\} \left[\exp\left\{ \int_t^{t+\tau} \epsilon(x) \ dx \right\} - 1\right] \frac{M(t+\tau)}{\eta(t)} \tag{2.4.10}$$

Introducing the notation $\bar{\epsilon}_{t,\tau}$ for $\dfrac{1}{\tau}\left[\int_t^{t+\tau} \epsilon(x) \ dx \right]$, and assuming small enough $\epsilon(x)$, (2.4.10) can be rewritten:

$$\epsilon(t) \simeq \frac{\tau R(t)}{\eta(t)} \bar{\epsilon}_{t,\tau} \tag{2.4.11}$$

Some general features of $\epsilon(t)$ are apparent from consideration of (2.4.11), although the exact solution depends on the conditions used to start the estimation process. We use the notation $\min_{x,y}\epsilon(t)$ for the minimum value of $\epsilon(t)$ over the the interval x to y and $\max_{x,y}\epsilon(t)$ for the equivalent maximum.

Consider an interval such that

$$\min_{t,t+\tau}\epsilon(t) > 0. \tag{2.4.12}$$

Over such an interval we know that

$$\min_{t,t+\tau}\epsilon(t) \leqslant \bar{\epsilon}_{t,\tau}$$

and (2.4.11) with (2.4.12) implies that

$$\min_{-\infty,t+\tau}\epsilon(t) > 0.$$

(2.4.11) and (2.4.12) also imply that if

$$\frac{\tau R(t)}{\eta(t)} > 1 \qquad (2.4.13)$$

then

$$\epsilon(t) > \min_{t, t+\tau} \epsilon(t). \qquad (2.4.14)$$

There is no condition equivalent to (2.4.14) involving $\max_{t, t+\tau} \epsilon(t)$ if (2.4.13) and (2.4.12) are true. A workable scheme of the general type (2.4.7) requires the estimation procedure to run in the direction of t decreasing, since recruitment must be calculated from *future* maturation. Hence once condition (2.4.12) has been satisfied for some interval of duration τ, $\epsilon(t)$ will tend to increase as t decreases over portions of a dataset for which (2.4.13) is true. In practice (2.4.13) is probably satisfied for the greater part of most of the real datasets to which our estimator might be applied: indeed in a population in which R, τ and η are constant $\tau R(t)$ is *never* less than $\eta(t)$.

A precisely analogous argument to the one presented above holds for the case

$$\max_{t, t+\tau} \epsilon(t) < 0$$

Again if (2.4.13) is true then the magnitude of $\epsilon(t)$ will tend to increase as t decreases.

To summarise, the mathematical form of the stage structure models (2.1.1) to (2.1.4) means that an estimator must be applied in the direction of increasing age and time or decreasing age and time but not, for example, decreasing age and increasing time. We have demonstrated that in the direction of increasing age and time the method is subject to age propagating instabilities and in the direction of decreasing age and time, time propagating instabilities come into play.

2.5. Discussion

The instabilities described in the previous section present fundamental obstacles to the development of time- and stage-dependent mortality estimators from any stage structured population models based on the same general assumptions as those embodied in equations (2.1.1)–(2.1.4). These are that within a stage all individuals develop at the same rate and have the same probability of death at any given time. We expect that mortality estimators derived from any modelling approach sharing these assumptions will be subject to similar instabilities. Age propagating instability appears then, somewhat regrettably, to be of fairly widespread significance in the attempt to estimate mortality from stage structured data. Time propagating instability blocks the most obvious way round the problem: working backwards rather than forwards through the stages.

What can be done to avoid these instabilities? Attempts by one of us (Wood 1989) to circumvent estimator instability by considering the population a few stages at a time and hence breaking the 'error cascade' failed to produce useful methods for inference of time- and stage-dependent *per capita* death rates, for every stage of a population. Nevertheless it is possible to obtain acceptable estimates of time dependent death rates for substantial subsections of a population using the method of section 2.3 (Wood *et al.* 1989). However this and similar efforts amount to efforts to salvage something from a fundamentally flawed approach. The key to further progress with the problem of inferring stage- and time-dependent death rates is to recognise the *biological* origin of the age propagating instability which is that the methods in this chapter make no assumption about smoothness of death rates *of neighbouring stages* at any given time. The validity of such an assumption depends on the underlying biology; it is more plausible, for example, for consecutive naupliar stages of a copepod than for larval and adult dragonflies.

The remainder of this monograph is concerned with methods applicable in situations where we expect some degree of continuity of death rates as a function of age/stage. In this work the stage structure formalism causes more problems than it resolves and we abandon it in favour of the more traditional description of a population in terms of its age structure.

CHAPTER 3

CUBIC SPLINES AND HISTOSPLINES

3.1 Introduction

In chapter 2 we provided an explanation for the failure of many methods intended to find time- and age-dependent mortality rates from stage structured population data. It was also suggested that an effective method of mortality estimation would be a more local method than those so far suggested. In chapter 4 we present such a method based on surface fitting with cubic spline functions. The method requires some standard techniques for fitting spline functions to noisy point data, which are outlined briefly in section 3.2. Chapter 4 also requires some extensions of spline theory to cover noisy aggregate data and covariance, which are the subjects of sections 3.3 and 3.4 respectively. Some of the development in section 3.3 is new, and section 3.5 briefly discusses some possible areas of application.

3.2 A brief guide to splines

The basic problem addressed by one-dimensional curve fitting is the construction of a function passing through or close to a set of points $\{(x_i, y_i)\}$. There is an infinite family of functions which will do this, but one member of this family has particularly useful geometric properties: the "cubic spline". The literature on splines is vast. For detailed information see Lancaster and Šalkauskas (1986) or de Boor (1978). Masochists may find Meir and Sharma (1972) more to their taste.

Cubic splines

Two 'points' or 'knots' (x_j, y_j), (x_{j+1}, y_{j+1}) and the gradients g_j and g_{j+1} required of a curve through these knots are sufficient to define a unique cubic polynomial passing through both points. A series of knots can therefore be interpolated by a set of piecewise cubic functions each joining two points in the series. Such an interpolant will be continuous and have a continuous first derivative. This idea is the basis for a sizeable family of interpolation schemes, since it provides a means for smoothly interpolating among a large number of points, whilst avoiding much of the spurious oscillation to which high order polynomial intepolation is prone (see chapter 2 of Lancaster and Šalkauskas, 1986).

If (as is commonly the case) the values of first derivatives at the knots are not available, then there are a number of schemes for estimating derivatives from function values (e.g. Yan 1987, Fritsch and Butland 1984). Probably the most successful of these is the method of cubic splines, which obtains first derivative values by assuming first *and* second derivative continuity (and requiring that two end conditions be supplied). The spline function interpolating

24

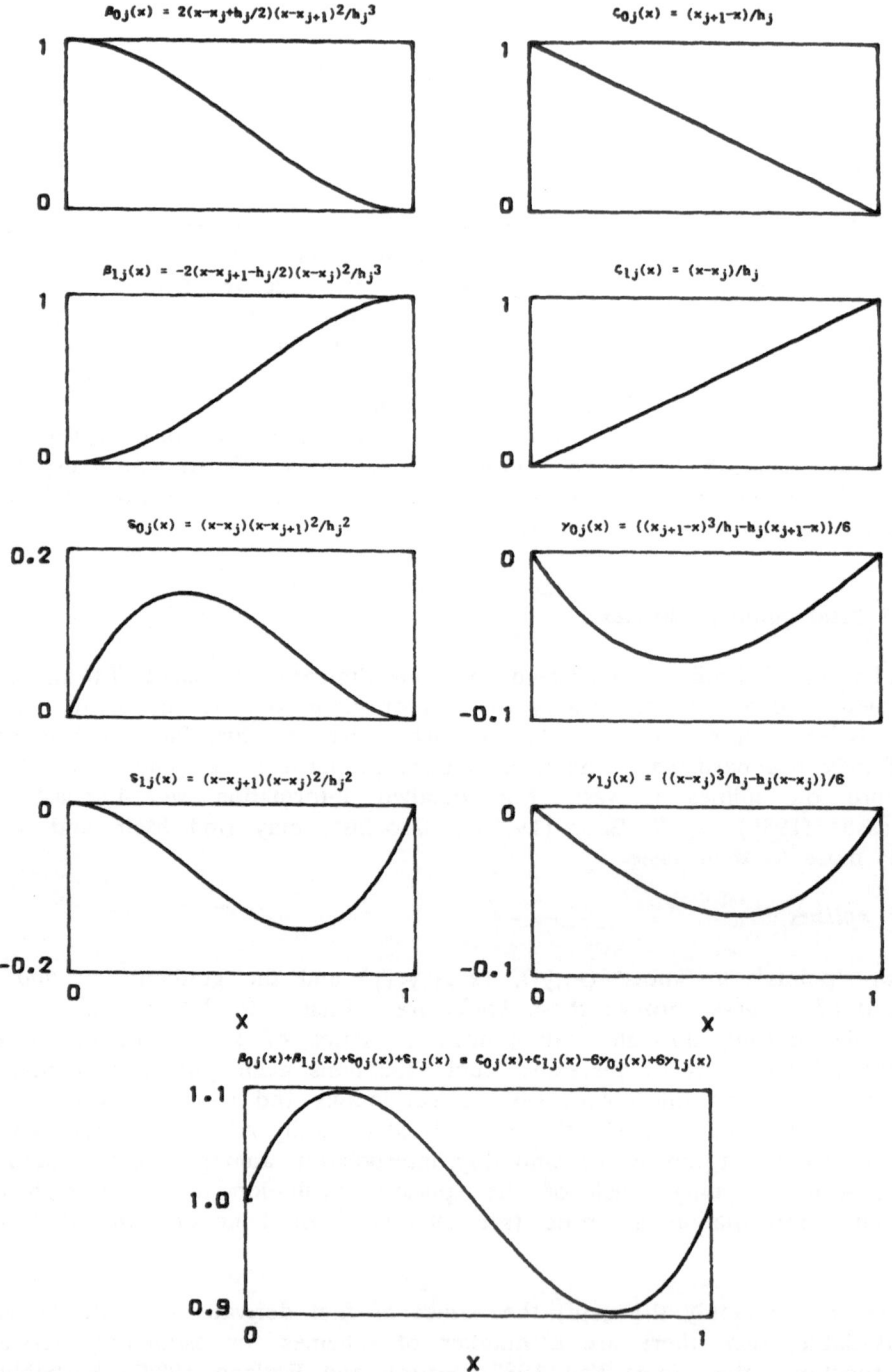

Figure 3.1 The basis functions for two alternative representations of a cubic spline function. The lower figure shows the weighted sum of either set of basis functions where the weights are chosen to yield a cubic with value and gradient of unity at both knots. Note that for this example $x_j = 0$, $x_{j+1} = 1$ and $h_j = x_{j+1} - x_j = 1$.

a set of knots (x_j, y_j) can be written

$$s(x) = y_j \beta_{0j}(x) + y_{j+1}\beta_{1j}(x) + g_j \delta_{0j}(x) + g_{j+1}\delta_{1j}(x), \quad x_j \leqslant x \leqslant x_{j+1} \qquad (3.2.1)$$

where β_{0j}, β_{1j}, δ_{0j} and δ_{1j} are cubic Hermite basis functions defined in figure 3.1. The continuity requirements for first and second derivatives yield a relationship between the gradients $\{g_j\}$ at the knots and the corresponding function values. This relationship is most conveniently expressed using matrix notation. Given a set of n knots, we define column vectors $\underline{y}$ and $\underline{g}$ by

$$\underline{y}^T - (y_1, y_2, \ldots \ldots y_n); \qquad \underline{g}^T - (g_1, g_2, \ldots \ldots g_n).$$

The two vectors are now related by a set of linear equations which we write in the shorthand form

$$T\underline{g} - D\underline{y}. \qquad (3.2.2)$$

where the elements of the matrices T and D corresponding to any given knot are listed in table 3.1, and are found by equating the second differentials at the knot of the cubic functions (3.2.1) on either side of the knot.

Alternatively, the spline can be written in terms of the vector of second derivatives, $\underline{m}$, at the knots. In this case

$$s(x) = y_j \varsigma_{0j}(x) + y_{j+1}\varsigma_{1j}(x) + m_j \gamma_{0j}(x) + m_{j+1}\gamma_{1j}(x) \quad x_j \leqslant x \leqslant x_{j+1} \qquad (3.2.3)$$

where the basis functions are again defined in figure 3.1. This time the elements of the matrices B and Q, relating $\underline{m}$ and $\underline{y}$, are obtained by equating the first derivatives at the knots of adjacent sections of cubic (3.2.3). This gives

$$B\underline{m} - Q\underline{y} \qquad (3.2.4)$$

where the elements of B, Q are defined in table 3.2. Note that the definitions given there are for a spline in which the second derivatives at the endpoints of the dataset are set to zero. The spline produced in this way is refered to as a 'natural' spline since it is the closest in shape to the draughtsperson's spline of which cubic spline functions are a close mathematical model.

The success of the spline approach lies in the fact that the cubic spline is the minimiser of the quantity

$$\int_{x_1}^{x_n} [s''(x)]^2 \, dx \qquad (3.2.5)$$

among all C^2 functions (functions with continuous first and second derivatives) passing through the knots (x_j, y_j) $1 \leqslant j \leqslant n$. This is convenient since this quantity is a reasonable measure of 'total curvature'.

Definitions of the matrices T and D

$$T_{j,j} = 2 \qquad\qquad j = 1, \ldots, k$$

$$T_{j,j-1} = h_j/(h_j + h_{j-1}) \qquad T_{j,j+1} = 1 - T_{j,j-1}$$

$$D_{j,j-1} = -3T_{j,j-1}/h_{j-1} \qquad D_{j,j+1} = 3T_{j,j+1}/h_j$$

$$D_{j,j} = -(D_{j,j+1} + D_{j,j-1}) \qquad\qquad j = 2, \ldots, k-1$$

End conditions	Corresponding Matrix Elements		
$s''(\alpha_0) = 0$	$T_{1,2} = 1$	$D_{1,1} = -3/h_1$	$D_{1,2} = -D_{1,1}$
$s''(\alpha_k) = 0$	$T_{k,k-1} = 1$	$D_{k,k-1} = -3/h_{k-1}$	$D_{k,k} = -D_{k,k-1}$
$s'(\alpha_0) = 0$	$T_{1,2} = 0$	$D_{1,1} = 0$	$D_{1,2} = 0$
$s'(\alpha_k) = 0$	$T_{k,k-1} = 0$	$D_{k,k-1} = 0$	$D_{k,k} = 0$

Definitions of the elements of matrices L and J

$L = Z\{\beta(x)\}$ and $J = Z\{\delta(x)\}$ where $Z\{\kappa(x)\}$ is defined as

$$Z_{j,j-1} = \int_{\alpha_{j-1}}^{a_j} \kappa_{00}(x)\,dx \qquad Z_{j,j+1} = \int_{x_j}^{\alpha_j} \kappa_{11}(x)\,dx$$

$$Z_{j,j} = \int_{\alpha_{j-1}}^{a_j} \kappa_{10}(x)\,dx + \int_{x_j}^{\alpha_j} \kappa_{01}(x)\,dx \qquad j = 1, \ldots, n$$

Columns zero and n+1 are only defined for the end conditions requiring both gradient and function value information.

Table 3.1 The definitions required to construct cubic splines using a basis of cubic Hermite polynomials and to construct area matching cubic splines. Note that $h_j = x_{j+1} - x_j$ and that all matrix elements not otherwise defined are zero.

The cubic smoothing spline

Introduced by Reinsch (1967), the cubic smoothing spline replaces exact interpolation with smoothing. Given a set of knots (x_j, a_j), instead of demanding that a curve pass through them all, we require that the spline $s(x)$ minimises the expression

$$\int_{x_1}^{x_n} [s''(x)]^2 \, dx \quad + \quad \rho \sum_{i=1}^{n} \{s(x_i) - a_i\}^2 w_i^2 \tag{3.2.6}$$

where w_i is a non-negative "weight" assigned to the i'th knot, and ρ is a positive constant (discussed later). Reinsch showed that for a given value of ρ, the coefficients of the smoothing spline, written in the form (3.2.3), come from the solution of

$$(QWQ^T + \rho B)\underline{m} = \rho Q\underline{a} \tag{3.2.7}$$

$$\underline{y} = \underline{a} - \rho^{-1}WQ^T\underline{m} \tag{3.2.8}$$

where $\underline{y}$ is the vector of values of $s(x_j)$, W is a diagonal matrix of weights w_i^2 and the elements of the matrices B and Q are defined in table 3.2. Note the distinction between y_i and a_i: the former is the value of the interpolating function when $x = x_i$, the latter is the y-coordinate of the knot.

Intuitively, equation (3.2.6) states that the cubic smoothing spline picks out a compromise between the opposing demands of minimising total curvature and minimising total (squared) deviation of the curve from the knots. The value of ρ defines the relative importance given to each demand. Note that as $\rho \to \infty$ $\underline{y} \to \underline{a}$, equation (3.2.7) reduces to equation (3.2.4), and an interpolating cubic spline is produced. Similarly as $\rho \to 0$, the requirement to minimise curvature is overriding, and the outcome is linear regression through the knots.

Following Hutchinson and de Hoog (1985) we can re-write this system as

$$\underline{y} = A_\rho \underline{a}$$

where A_ρ is defined by

$$I - A_\rho = WQ \, Z_\rho^{-1} \, Q^T \tag{3.2.9}$$

and

$$Z_\rho = (QWQ^T + \rho B).$$

We can also write

$$\underline{m} = C_\rho \underline{a}$$

defining

$$C_\rho = B^{-1}QA_\rho. \tag{3.2.10}$$

Definitions of the matrices B and Q

$$Q_{j,j} = \frac{1}{h_j} \qquad Q_{j,j+1} = -\left(\frac{1}{h_j} + \frac{1}{h_{j+1}}\right)$$

$$Q_{j,j+2} = \frac{1}{h_{j+1}} \qquad B_{j,j} = \frac{1}{3}(h_j + h_{j+1}) \qquad j = 1, \ldots, n-2$$

$$B_{j,j+1} = \frac{1}{6}h_{j+1} \quad B_{j+1,j} = \frac{1}{6}h_{j+1} \qquad j = 1, \ldots, n-3$$

Definitions of the elements of matrices R and P

$$R_{j,j} = \int_{x_j}^{\alpha_j} \varsigma_{1j}(x)\,dx \qquad\qquad j = 1, \ldots, n-2$$

$$R_{j,j-1} = \int_{x_j}^{\alpha_j} \varsigma_{0j}(x)\,dx + \int_{\alpha_{j-1}}^{x_j} \varsigma_{1j-1}(x)\,dx \qquad j = 2, \ldots, n-1$$

$$R_{j,j-2} = \int_{\alpha_{j-1}}^{x_j} \varsigma_{0j-1}(x)\,dx \qquad\qquad j = 3, \ldots, n$$

$$P_{j,j+1} = \int_{x_j}^{\alpha_j} \gamma_{1j}(x)\,dx \qquad\qquad j = 1, \ldots, n-1$$

$$P_{j,j} = \int_{x_j}^{\alpha_j} \gamma_{0j}(x)\,dx + \int_{\alpha_{j-1}}^{x_j} \gamma_{1j-1}(x)\,dx \qquad j = 2, \ldots, n-1$$

$$P_{j,j-1} = \int_{\alpha_{j-1}}^{x_j} \gamma_{0j-1}(x)\,dx \qquad\qquad j = 2, \ldots, n$$

$$P_{1,1} = \int_{\alpha_0}^{\alpha_1} \gamma_{01}(x)\,dx \qquad P_{n,n} = \int_{\alpha_{n-1}}^{\alpha_n} \gamma_{1n-1}(x)\,dx$$

Table 3.2 Matrix definitions required for a spline described in terms of second derivatives at its knots and to construct area approximating splines. Note that $h_j = x_{j+1} - x_j$ and that all matrix elements not otherwise defined are zero.

Choosing the smoothing parameter, ρ

Reinsch (1967) originally suggested that the smoothing parameter ρ should be chosen so that the mean residual sum of squares equals the expected variance, σ^2, of the error affecting the data, if this is known. Wahba and Wold (1975) point out that this approach leads to significant over–smoothing and a better solution follows from assuming the error model $E\{\epsilon_j\}=0$, $E\{\epsilon_i\epsilon_j\}=0$ for $i \neq j$ and $E\{\epsilon_j{}^2\}=w_j{}^2\sigma^2$ in which $E\{\ \}$ denotes expectation value. If σ^2 is known then it can be shown that minimisation of the function $V(\rho)$ defined by

$$V(\rho) = \frac{1}{n} \|W^{-1}(I - A_\rho)\underline{y}\|^2 - \frac{2\sigma^2}{n} Tr(I - A_\rho) + \sigma^2 \qquad (3.2.11)$$

with respect to ρ (where $\|\ \|$ denotes Euclidean norm), is equivalent to minimising the sum of squares of the expected deviation of the spline function at the knots from the function which it is supposed to approximate. In section 3.4 we extend this standard result to cover the situation where there is covariance in the error terms (i.e. $E(\epsilon_i\epsilon_j) \neq 0$, when $i \neq j$).

If σ^2 is unknown, but we can introduce the further assumption of a normal error distribution, then Craven and Wahba (1979) show that ρ can be found by minimising

$$G_{cv}(\rho) = \frac{n\|W^{-1}(I - A_\rho)\underline{a}\|^2}{\{Tr(I - A_\rho)\}^2} \qquad (3.2.12)$$

As $n \to \infty$ the value of ρ calculated by minimising $G_{cv}(\rho)$ tends to the value given by the minimisation of $V(\rho)$. Intuitively G_{cv} is a measure of the average deviation of a given spline function from a data point not used in the calculation of its coefficients; the technique is therefore given the name 'cross validation'.

Wahba (1983) showed how to obtain confidence intervals (CI) such that there is a 95% probability that the "true" line lies within $\pm$ 2CI of the spline, for a spline fitted by cross validation. She obtained estimates of σ^2 for the original data:

$$\hat{\sigma}^2 = \frac{\|W^{-1}(I - A_\rho)a^2\|}{Tr(I - A)_\rho} \qquad (3.2.13)$$

and

$$CI = 2\hat{\sigma}^2 / (Tr(A_\rho)/n). \qquad (3.2.14)$$

She also suggested after extensive Monte Carlo experiments that cross validation produces unreliable results for fewer than 30 datapoints, a point we develop further in section 4.6. Ragozin (1983) has produced error bounds for the smoothing spline which also take into account the expected deviation of the spline from the underlying function independent of random sampling error, but to use these results we would need to have bounds on the second derivative of the underlying function.

Practical calculation of smoothing splines

The matrices in the linear systems (3.2.7) and (3.2.8) are at most five banded, making their solution simple and efficient if ρ is predetermined: Reinsch (1967) suggested an algorithm based on Choleski decomposition. If ρ is unknown then matters are more complicated. To choose ρ by minimisation of (3.2.11) or (3.2.12), requires evaluation of the terms involving A_ρ for a number of values of ρ. Craven and Wahba (1979) opted for the full frontal approach to this problem and calculated the elements of A_ρ directly at considerable computational expence. Utreras (1985) and Silverman (1984) then suggested less computationally thirsty approximate methods for evaluating (3.2.11) and (3.2.12) before Hutchinson and de Hoog (1985) developed an efficient exact algorithm (for Fortran code see Hutchinson, 1986). This method although very fast suffers from rounding error problems for high ρ values and the same authors later adapted the work of Elden (1984) to produce a slightly slower but very robust algorithm based on Givens rotations (de Hoog and Hutchinson, 1987). It is this latter algorithm that we use for calculating smoothing splines.

3.3 Cubic splines for exact or noisy histogram data

Although spline functions applicable to point data have been extensively studied, the related problems for aggregated data have received less attention. Consider a distribution which is adequately described by a continuous function, $f(x)$, over the interval $x=\alpha_0$ to $x=\alpha_n$. This distribution is sampled, yielding data of the form:

$$\eta_j = \int_{\alpha_{j-1}}^{\alpha_j} f(x) \, dx + \epsilon_j, \quad \text{for } j = 1,..,n, \tag{3.3.1}$$

where ϵ_j is a normally distributed error term characterised by the covariance matrix, Co. Given the vector of measurements, η and an estimate of Co we wish to obtain an estimate of $f(x)$.

Since the work of Boneva, Kendall and Stefanov (1972) the use of area-matching spline functions has been recognised as a way of approximating the distribution underlying histogram data, which avoids the subjective element involved in the choice of a parametric function to describe the distribution. The disadvantages of their approach are that it deals only with exact data and relies on quadratic and quartic splines, so failing to make use of the well known geometric properties (equation 3.2.5) of cubic splines (see, for example, de Boor 1978). In the next subsection a cubic area matching spline is presented and the positivity constraint on the distribution estimate is discussed.

Dyn and Wahba (1982) and Wahba (1981) have dealt with the problem of distribution estimation from aggregated noisy data in the bivariate case. Their methods are based on thin plate splines, but for univariate data cubic smoothing splines are a simpler way of tackling the problem and these are the subject of the following subsection.

A cubic area matching spline

A cubic spline, s, interpolating a set knots (x_j, y_j) can be defined by equations (3.2.1) and (3.2.2). For aggregated data the positions of the knots, x_j, are to a certain extent arbitrary, but in what follows we use interval midpoints; thus we define $x_j = (\alpha_j + \alpha_{j-1})/2$ for $j=2,..,n-1$. The number of knots at the ends of the spline is determined by the type of end condition used. Two sets of end conditions are of interest here: the case in which the second derivative of the spline is set to zero at one or both ends (the so called 'natural' end condition), and the case in which both the value and gradient of the spline are specified at an end. For a 'natural' spline at α_0, we set $x_1 = \alpha_0$; otherwise $x_0 = \alpha_0$ and $x_1 = (\alpha_1 + \alpha_0)/2$. The equivalent definitions at α_n are $x_n = \alpha_n$ or $x_n = (\alpha_{n-1} + \alpha_n)/2$ and $x_{n+1} = \alpha_n$ respectively.

Defining

$$\eta_j^* = \int_{\alpha_{j-1}}^{\alpha_j} s(x) \, dx,$$

it is clear from integrating (3.2.1) that the vector $\underline{\eta}^* = (\eta_1, \ldots, \eta_n)^T$ is related to the vectors $\underline{y}$ and g by the equation

$$\underline{\eta}^* = J\underline{y} + Lg$$

where L and J are defined in table 3.1. Using (3.2.2) this can be written

$$\underline{\eta}^* = \Lambda\underline{y} \quad \text{where} \quad \Lambda = J + LT^{-1}D.$$

For a natural spline we set $\underline{\eta} = \underline{\eta}^*$, but if both function values and gradients at the endpoints are to be supplied then we must amend the system so that

$$\begin{bmatrix} y_0 \\ \underline{\eta} \\ y_{n+1} \end{bmatrix} = \begin{bmatrix} 1, \ldots \ldots, 0 \\ \Lambda \\ 0, \ldots \ldots, 1 \end{bmatrix} \underline{y}. \qquad (3.3.2)$$

It is of course possible to mix one type of end condition with another. Solution of (3.3.2) or one of its relations for y and (3.2.2) for g completely specifies a cubic spline matching each value η_j exactly.

The distribution estimates produced by the above method are not necessarily positively constrained. However, area–matching cubic splines calculated in this way make it easy to produce simple algorithms for the production of positively constrained distribution estimates with continuous first derivative. Figure 3.3 shows an example produced using the algorithm outlined in the appendix.

Although applicable in the study of human demography and in some areas of epidemiology where accurate data is available, the area matching schemes presented here are of limited value in the noise ridden fields of ecology and population biology (but see section 4.7 for an exception). We now develop a method for reconstructing continuous distributions from noisy histogram data.

An area approximating smoothing spline

We seek a 'natural' spline function approximating f(x) in equation (3.3.1) in the case when the error terms ϵ_j are significant. Presentation of this smoothing spline is facilitated by using the spline defined by equations (3.2.3) and (3.2.4), thus working with the second derivatives of the spline at the knots. Using this representation

$$\eta^* = (P + RB^{-1}Q)y,$$

where the matrices B and Q are introduced in (3.2.4) and elements of matrices P and Q are calculated from integration of (3.2.3): the elements of all four matrices are given in table 3.2.

To find the area approximating cubic smoothing spline we must minimise

$$\lambda(\rho) = \int_{\alpha_0}^{\alpha_n} [s''(x)]^2 dx + \rho(\eta^* - \eta)^T Co^{-1}(\eta^* - \eta) \qquad (3.3.3)$$

where ρ is a parameter controlling the trade off between fidelity to the data and 'smoothness' of the spline function. This expression replaces (3.2.6) in section (3.2.2). Since s(x) is a *cubic* spline s''(x) is a piecewise linear function, hence the first term of $\lambda(\rho)$ can be expressed in terms of the basis of linear spline functions as described, for example, in Lancaster & Salkauskas (1986). The second term is a quadratic form. It follows that minimisation of $\lambda(\rho)$ with respect to y (i.e. with ρ fixed for the moment) requires the solution of the linear system

$$\Xi y + \rho \Psi^T Co^{-1}(\Psi y - \eta) = 0$$

where 0 is the zero vector, $\Psi = P + RB^{-1}Q$ and $\Xi = Q^T B^{-1}Q$. Defining

$\Gamma = \Psi^T Co^{-1} \Psi$ it is trivial to write

$$\underline{y} = Y_\rho \underline{\eta} \quad \text{where} \quad Y_\rho = [\Xi + \rho\Gamma]^{-1} \rho \Psi^T Co^{-1} \tag{3.3.4}$$

and

$$\underline{m} = M_\rho \underline{\eta} \quad \text{where} \quad M_\rho = B^{-1} Q[\Xi + \rho\Gamma]^{-1} \rho \Gamma^T Co^{-1}. \tag{3.3.5}$$

It remains to establish criteria for choosing the smoothing parameter ρ. To this end we define the 'influence matrix' A_ρ for the system, such that $\underline{\eta}^* = A_\rho \underline{\eta}$. Explicitly this means that

$$A_\rho = \Psi[\Xi + \rho\Gamma]^{-1} \rho \Psi^T Co^{-1}.$$

In the absence of covariance between the error terms in (3.3.1) the smoothing parameter can be chosen by minimisation of (3.2.11) or (3.2.12) after replacing $\underline{a}$ by $\underline{\eta}$, but if an estimate of the covariance matrix for the data is available then the results of the next section can be used.

3.4 Choosing the smoothing parameter with covariant error terms

If the errors affecting a dataset are independent then equation (3.2.11) can be used to choose ρ, but this equation must be extended if it is to be applicable to a covariant error model. Rewriting (3.3.1) as $\eta_j = \eta_{tj} + \epsilon_j$ standard least squares theory dictates that ρ should be chosen to minimise

$$T(\rho) = (\underline{\eta}^* - \underline{\eta}_t)^T Co^{-1} (\underline{\eta}^* - \underline{\eta}_t)/n,$$

(see, for example Meyer, 1975). Since the vector of true integrals $\underline{\eta}_t$ is unknown we cannot evaluate $T(\rho)$ directly but must deal instead with its expected value. Since Co is symmetric we can form the Choleski decomposition of Co^{-1}, LL^T, and write

$$E\{T(\rho)\} = E\{\|L^T(A_\rho \underline{\eta} - \underline{\eta}_t)\|^2/n.\}$$

Under the assumption that $\underline{\eta}_t$ and $\underline{\epsilon}$ are independent this can be recast

$$E\{T(\rho)\} = E\{\frac{1}{n} \|L^T(A_\rho - I)\underline{\eta}_t\|^2\} + \frac{1}{n} \sum_i \sum_j \sum_k (L^T A_\rho)_{ki} (L^T A_\rho)_{kj} Co_{ij}$$

where the summations are from 1 to n. Row by row expansion of the first term in this expression along with an additional substitution of $\underline{\eta}_t + \underline{\epsilon}$ for $\underline{\eta}$ leads to

$$E\{T(\rho)\} ~-~ \frac{1}{n}~\|L^T(I-A_\rho)\underline{\eta}\|^2 ~+~ \frac{1}{n}~\underset{i~j~k}{\Sigma~\Sigma~\Sigma}~(L^TA_\rho)_{ki}(L^TA_\rho)_{kj}Co_{ij}$$

$$-~\frac{1}{n}~\underset{i~j~k}{\Sigma~\Sigma~\Sigma}~(L^T(I-A_\rho))_{ki}(L^T(I-A_\rho))_{kj}Co_{ij}$$

which after a few more lines of unremitting tedium leads to the function $V(\rho)=E\{T(\rho)\}$ which should be minimised to find ρ:

$$V(\rho)-~\frac{1}{n}\|L^T(I-A_\rho)\underline{\eta}\|^2+~\frac{2}{n}~\underset{i~j}{\Sigma~\Sigma}(Co^{-1}A_\rho)_{ij}Co_{ij}-~\frac{1}{n}~\underset{i~j}{\Sigma~\Sigma}~Co^{-1}{}_{ij}Co_{ij}~~~(3.4.1)$$

The zealous may like to check that this does indeed reduce to (3.2.11) for independent errors.

3.5 Two examples of the application of cubic area splines to age–structured population data

Using cubic area matching splines to estimate the distribution of Scottish women infected with HIV

Age distributions are inherently non–negative and so any continuous distribution estimate based on histogram data should share this property. The algorithm in the appendix uses the cubic area matching splines of section 3.3.1 to produce a C^1 non–negative area–matching curve through histogram data. Figure 3.2A shows this algorithm applied to data on the number of Scottish women known to be HIV–antibody positive. The data is from the Scottish AIDS Monitor (dated: May 1989). For comparison figure 3.2B shows an area–matching quadratic spline applied to the same data. The cubic scheme provides a non–negative distribution estimate which is in this case certainly not less smooth than the results using the quadratic spline.

Comparison of the area–approximating spline with two simpler distribution estimation schemes

Hay *et.al.* (1988) proposed a model of copepod population dynamics which, if we assume that *per capita* mortality rate is time–independent, implies an age structure at any time represented by the sum of two Gaussians multiplied by a negative exponential term. We therefore chose to test area–approximating splines on the model age structure function:

$$f(\alpha)~-~[exp\{-(2-\alpha)^2/4\}~+~2exp\{-(9-\alpha)^2/8\}]exp(-\alpha/5)$$

This function was sampled by numerical integration over n contiguous age classes α_{i-1} to α_i where $0 \leqslant i \leqslant n$. For $0 < i < n$, α_i was chosen randomly such that if we define $\delta\alpha = (\alpha_n - \alpha_0)/[3(n-1)]$ and $\Delta\alpha = (\alpha_n - \alpha_0 - \delta\alpha)/(n-1)$ then α_i had uniform probability of lying anywhere in the interval $(i-1)\Delta\alpha + \delta\alpha$ to $i\Delta\alpha$. $\alpha_0 = 0$ and $\alpha_n = 10$. To test the method, datasets were constructed with n running from 7 to 19 in steps of 2. This simulated sample data was then perturbed by

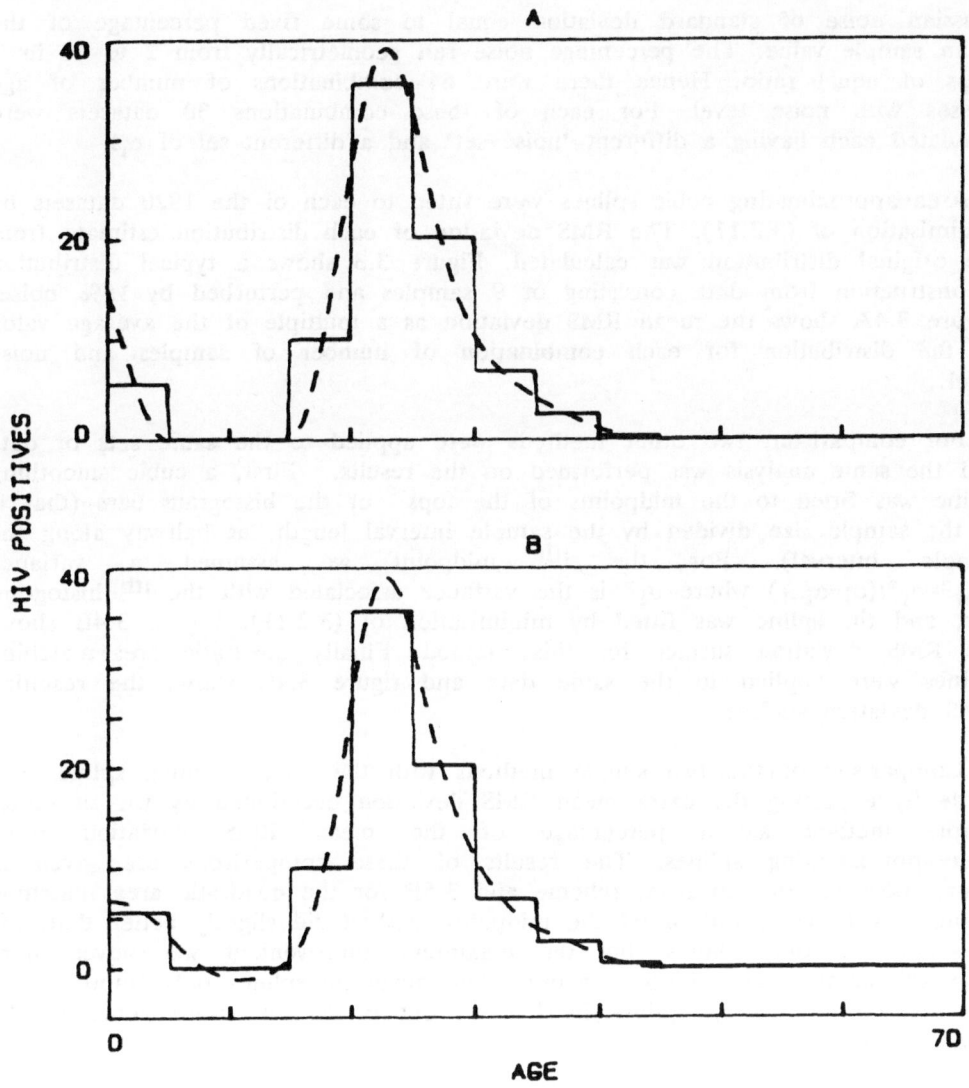

Figure 3.2 Spline based estimates of the age distribution of Scottish women known to be HIV positive. Figure A shows an estimate constrained to be positive by the algorithm of the appendix using cubic area-matching splines. Figure B shows an estimate made using quadratic histosplines. In both figures the dashed line is the distribution estimate and the continuous line the histogram data from which it is calculated.

gaussian noise of standard deviation equal to some fixed percentage of the mean sample value. The percentage noise ran geometrically from 2 to 40 in 7 steps of equal ratio. Hence there were 64 combinations of number of age classes with noise level. For each of these combinations 30 datasets were simulated each having a different 'noise–set' and a different set of α_is.

Area–approximating cubic splines were fitted to each of the 1920 datasets by minimisation of (3.2.11). The RMS deviation of each distribution estimate from the original distribution was calculated. Figure 3.3 shows a typical distribution reconstruction from data consisting of 9 samples and perturbed by 15% noise. Figure 3.4A shows the mean RMS deviation as a multiple of the average value of the distribution for each combination of number of samples and noise level.

For comparison, two other methods were applied to the same sets of data and the same analysis was performed on the results. First, a cubic smoothing spline was fitted to the midpoints of the tops of the histogram bars (that is to the sample size divided by the sample interval length, at halfway along the sample interval). For the ith midpoint we assumed a variance $\sigma_{mi}^2 = \sigma_i^2 / (\alpha_i - \alpha_{i-1})$ where σ_i^2 is the variance associated with the ith histogram bar, and the spline was fitted by minimisation of (3.2.11). Figure 3.4B shows the RMS deviation surface for this method. Finally quadratic area–matching splines were applied to the same data and figure 3.4C shows the resulting RMS deviation surface.

Comparison of the two simple methods with the area–matching splines was made by expessing the extra mean RMS deviation occasioned by use of either simple method as a percentage of the mean RMS deviation using area–approximating splines. The results of these comparisons are given in figure 3.5A for the midpoint scheme and 3.5B for the quadratic area matching splines. In 17 cases out of 64 the midpoint method did slightly better than the area–approximating splines, but the maximum improvement was never more than 4% of the mean RMS deviation. The quadratic splines only improved on the area approximating splines in 2 cases out of 64 and were generally the worst method.

It seems from this example that area–approximating splines are worth using when there are few samples and/or the data is not very noisy (at least by the standards of experimental ecology). For very noisy data the infelicities of the midpoint method are insignificant in comparison to the random noise and with increasingly finely sampled data the approximations of the midpoint method become increasingly accurate.

Copepod age structure data typically consist of some 9 to 11 age classes, or even fewer if nauplii are not distinguished into classes. It is therefore likely that area–approximating splines will better approximate copepod age structures than simpler schemes, and are in any case unlikely to do worse.

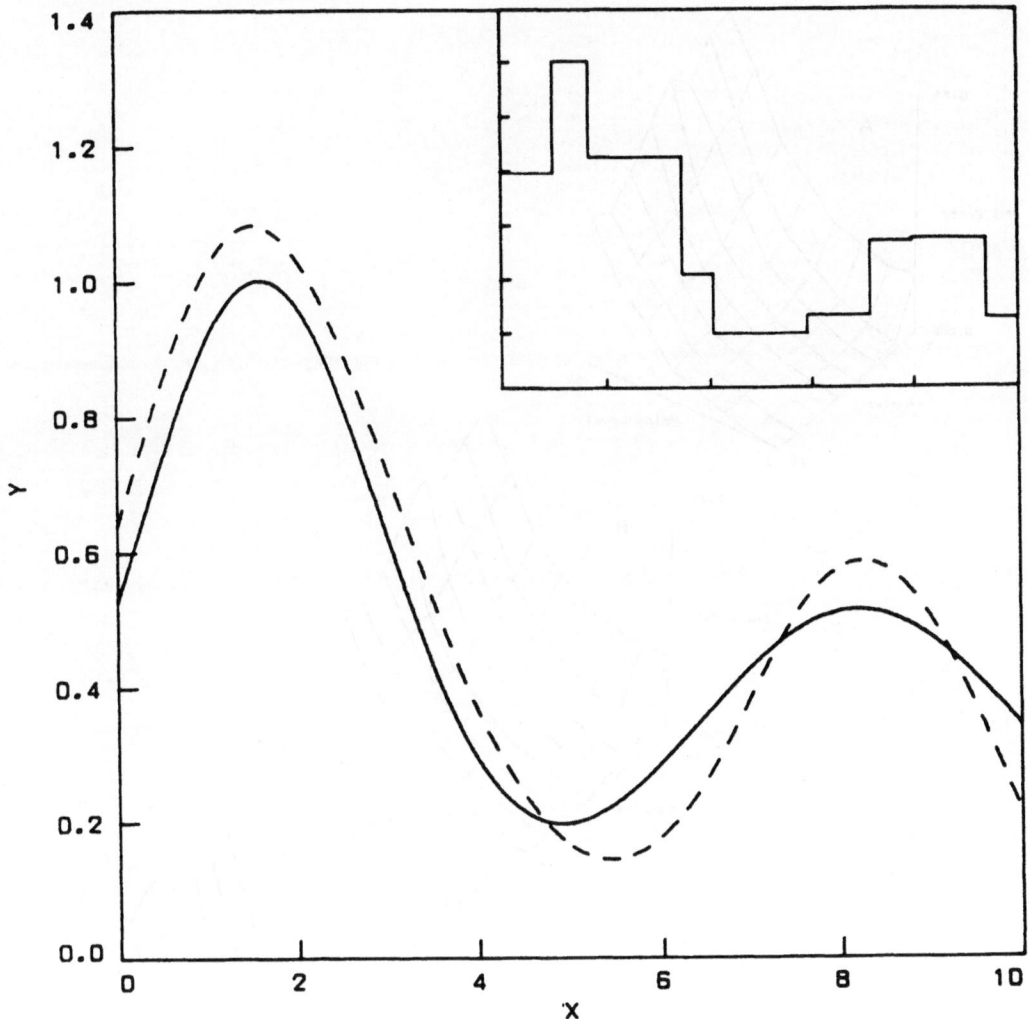

Figure 3.3 The cubic area—approximating spline applied to simulated age—structure data. The continuous line is the original distribution, the dashed line the reconstruction. The insert shows the noisy histogram data to which the method was applied. In this example the data was peturbed by gaussian noise with standard deviation 15% of the mean historgam bar area.

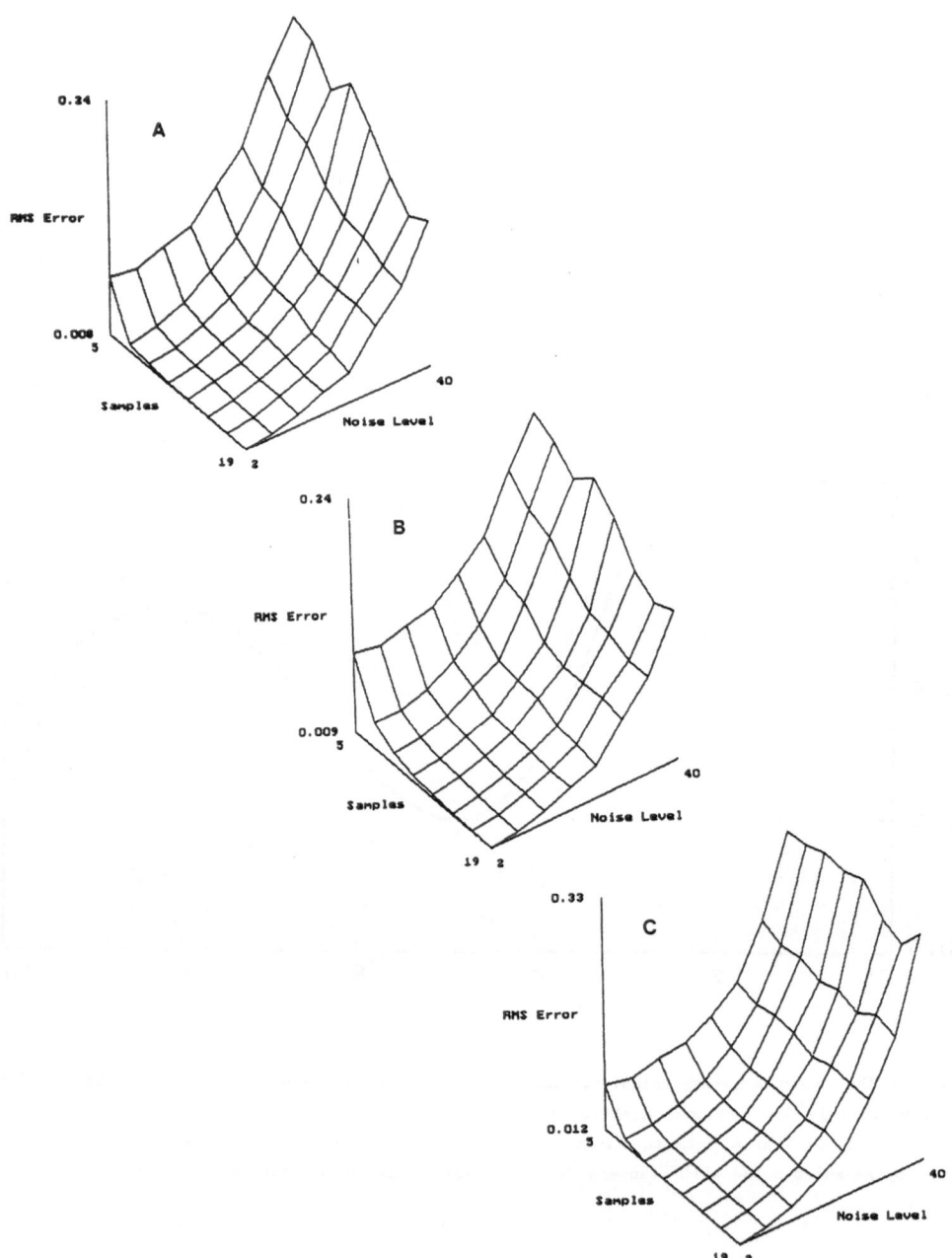

Figure 3.4 RMS deviations of reconstructed distributions from the original simulated distribution are given as multiples of the mean value of the original distribution. The deviations are plotted as functions of the number of histogram bars used to reconstruct the distribution (linear scale) and the level of noise applied to the histogram data as a percentage of the mean of the original distribution (log scale). A) for an area-approximating spline, B) for the simple smoothing scheme described in the text and C) for a quadratic area-matching spline.

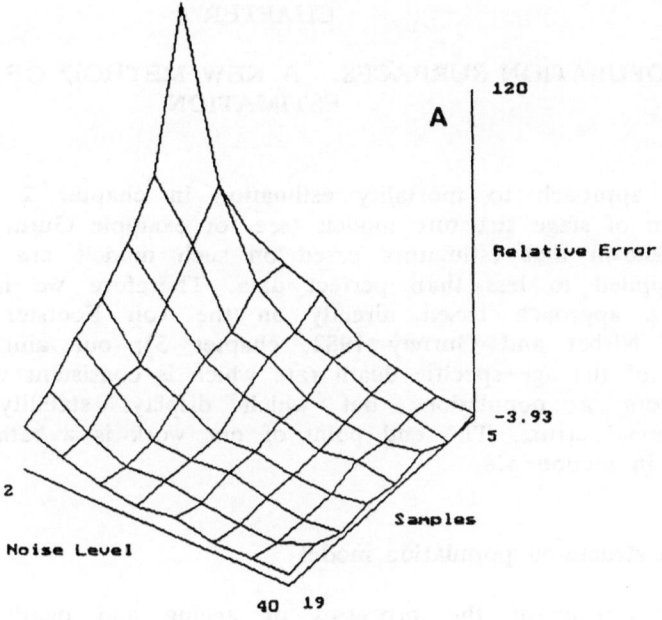

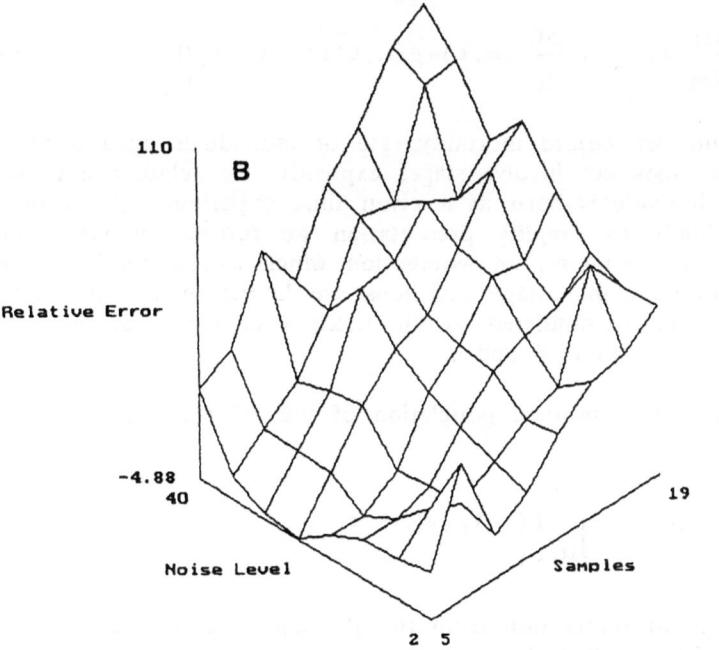

Figure 3.5 The mean additional error in comparison to the area-approximating spline introduced by using: A) the simple smoothing scheme and B) the quadratic histospline. The additional error is given as a percentage of the error produced by using the area-approximating spline. The remaining axes are the same as those for Figure 3.4.

CHAPTER 4

POPULATION SURFACES: A NEW METHOD OF MORTALITY ESTIMATION

The approach to mortality estimation in chapter 2 was based on the formalism of stage structure models (see for example Gurney *et. al.* 1983), but it was shown that estimators based on such models are prone to instability when applied to less than perfect data. Therefore we now consider a less restrictive approach based directly on the von Foerster equation (see for example Nisbet and Gurney 1982, chapter 3); our aim is to produce an estimate of the age–specific death rate which is consistent with stage abundance data from a population, but which displays stability with respect to observational errors. The end point of our work is a rather elaborate recipe, detailed in section 4.8.

4.1 The structured population model

The effects of the processes of ageing and death on an age– and time–dependent population density $f(\alpha,t)$ can be described by the von Foerster equation:

$$\frac{\partial f}{\partial \alpha}(\alpha,t) + \frac{\partial f}{\partial t}(\alpha,t) + \mu(\alpha,t)f(\alpha,t) = 0 \qquad (4.1.1)$$

where $\mu(\alpha,t)$ is the *per capita* mortality rate of individuals aged α at time t. Since this equation does not involve stages explicitly, we relate age to stage by assuming that all individuals born at a given time experience the same *known* stage durations. Solely to simplify presentation we further suppose that stage durations are time–independent, an assumption which can easily be relaxed. In practice the requirement that stage durations are known means that laboratory measurements must be extrapolated to the field: a controversial process which we discuss further in chapters 6 and 7.

Using this model the observed population of the jth stage at the ith sample time t_i is

$$P_{j,i} = \int_{\alpha_{j-1}}^{\alpha_j} f(x,t_i)\,dx + \epsilon_{j,i} \qquad (4.1.2)$$

where α_j is the age at maturation from the jth stage and $\epsilon_{j,i}$ is an error term with as yet unspecified statistical properties.

Our aim is to use the von Foerster equation to infer death rates from stage population time series data without encountering the instabilities identified in chapter 2. Since the arguments in that chapter relating to age propagating instability are very general, some further constraint on admissable forms for the death rate is clearly necessary. We therefore assume *smoothness with respect to age and time* thereby restricting applicability of the resulting method to situations where individuals in successive stages occupy similar niches. The method is thus applicable to zooplankton but not to Scottish highland midges (whose juveniles while away their time innocently consuming plankton whilst the adults terrorise anything with the capacity to itch).

Mathematically, we are recognising that there is a large family of alternative per capita death rate functions, consistent with (4.1.1) and with any specified set of stage populations $P_{j,i}$. By using spline functions to estimate $f(\alpha,t)$ we can turn this redundancy to our advantage by assuming that the surface is smooth in the sense implied by the minimisation of functions like (3.2.5), (3.2.6) or (3.3.3), thereby eliminating the instabilities which have plagued previous methods.

4.2 Preliminary estimation of the population surface $f(\alpha,t)$

The population $\eta_j(t)$, of any stage j at time t, can be estimated by using generalised cross validation to fit a cubic smoothing spline to the raw sample data $\underline{p}_j$ where p_{ji} is the population of the jth stage at the ith sample time t_i. Figure 4.1 shows an example using data on a copepod population in an enclosure experiment by J. Gamble and S. Hay (see chapter 6). Using the matrices A_ρ and C_ρ defined by equations (3.2.9) and (3.2.10) and equation (3.2.3) for a cubic spline we can write

$$\eta_j(t) = \underline{k}_j(t).\underline{p}_j$$

where $\quad \underline{k}_j(t) = \varsigma_{0i}(t)\underline{A}_{ji} + \varsigma_{1i}(t)\underline{A}_{ji+1} + \gamma_{0i}(t)\underline{C}_{ji} + \gamma_{1i}(t)\underline{C}_{ji+1}$,

"." represents a scalar product and the notation $\underline{C}_{ji}$ ($\underline{A}_{ji}$) is introduced for the ith row of the matrix C_ρ (A_ρ) associated with the jth stage. Note that these quantities are *column vectors*. Given the error model of section 3.2, equation (3.2.13) can be used to estimate the vector whose components are the variances associated with each datapoint

$$\underline{\xi}_j = \left[w^2_{j_1}\hat{\sigma}^2, w^2_{j_2}\hat{\sigma}^2, \quad . \quad . \quad . \quad , w^2_{j_n}\hat{\sigma}^2 \right]^T. \qquad (4.2.1)$$

We can now estimate $f(\alpha,t)$ by applying area approximating or area matching splines to the aggregated age structure estimates at time t: $\underline{\eta}(t)=[\eta_1(t),\eta_2(t), \quad . \quad . \quad . \quad ,\eta_e(t)]^T$ where e is the number of pre-adult stages (see figure 4.2). Defining

$$\underline{p} = (\underline{p}^T_1, \underline{p}^T_2, \quad . \quad . \quad . \quad ,\underline{p}^T_e)^T, \qquad (4.2.2)$$

42

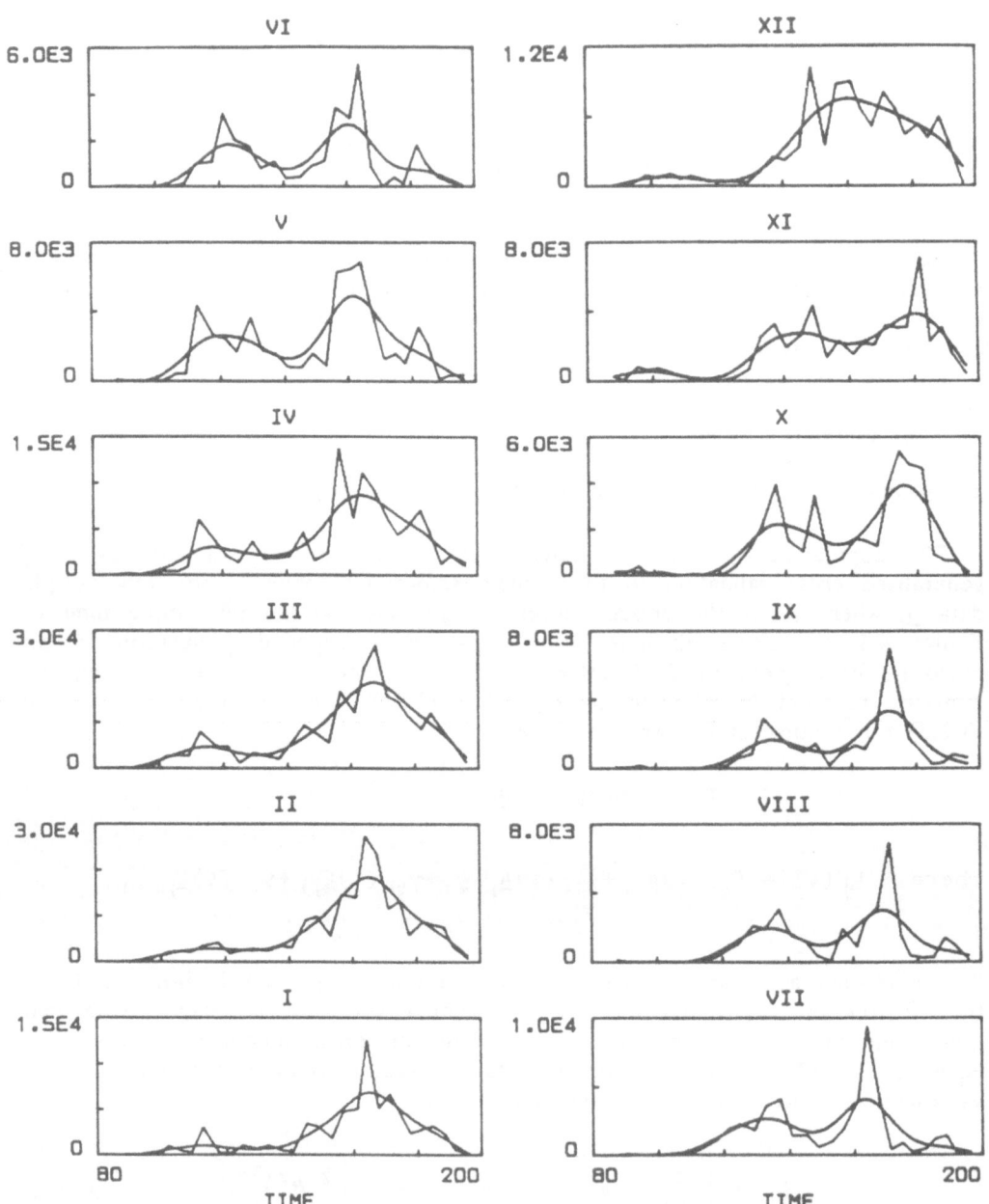

Figure 4.1 Cubic smoothing splines fitted to the 12 stages of a copepod population (*Pseudocalanus* 1980 Loch Ewe, Scotland) by 'generalized cross validation'. The smoothing parameter was chosen as described in section 4.6. The piecewise linear line joins the original data points and the smooth line is the cubic smoothing spline.

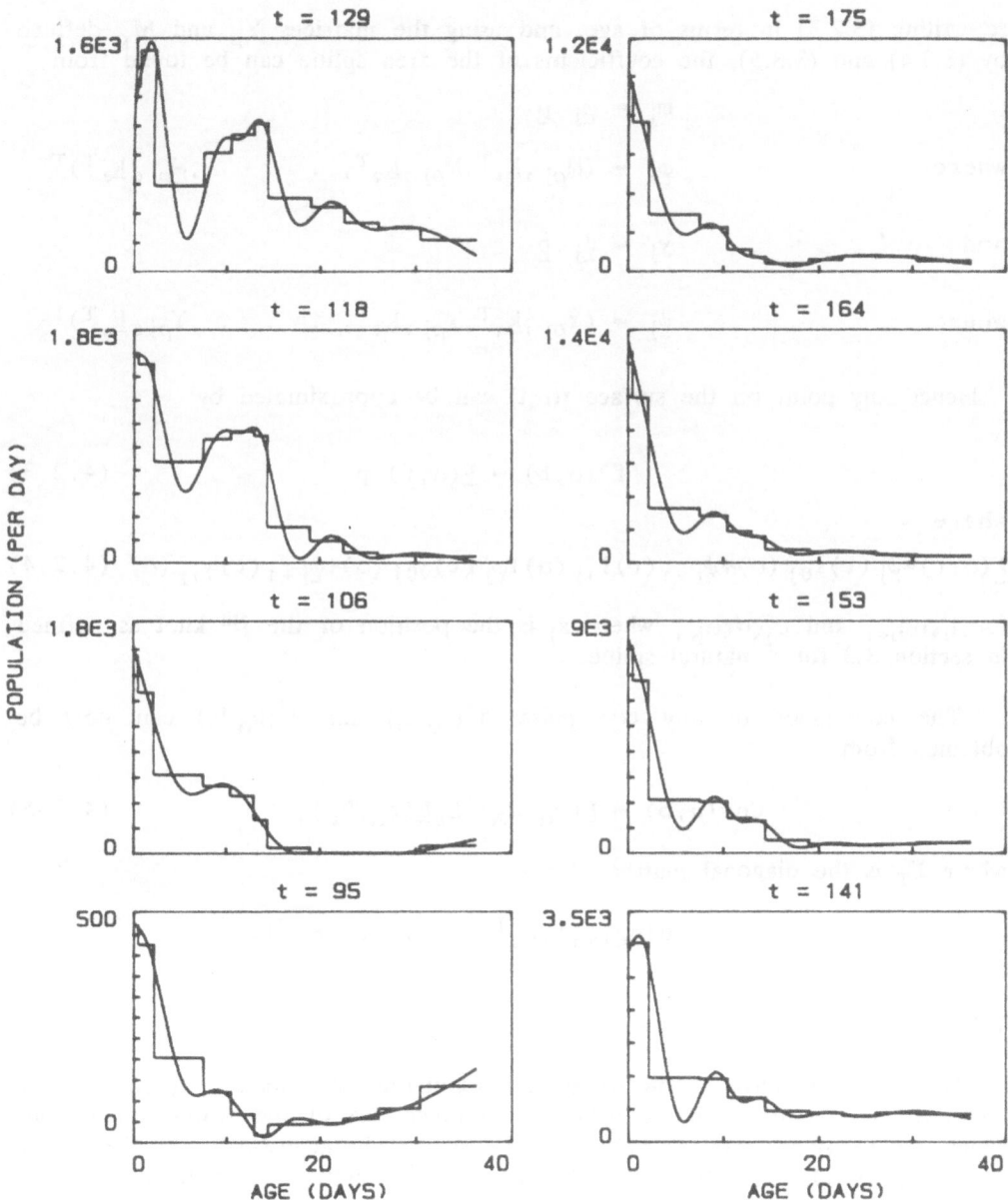

Figure 4.2 Area-matching splines fitted to a series of age structure histograms for the dataset used in figure 4.1, to obtain estimates of the continuous age distribution at each time. The initial estimate of $f(\alpha, t)$ derived in section 4.2 is essentially a time series of such age distribution estimates.

re-writing (3.2.3) in terms of age, and using the matrices Y_ρ and M_ρ defined by (3.3.4) and (3.3.5), the coefficients of the area spline can be found from

$$m_j = \underline{\varphi}_j \cdot \underline{p}$$

where

$$\underline{\varphi}_j = (M_{\rho j\,1} \underline{k}_1{}^T, M_{\rho j\,2} \underline{k}_2{}^T, \quad . \quad . \quad . \quad , M_{\rho j\,e} \underline{k}_e{}^T)^T$$

and

$$y_j = \underline{\psi}_j \cdot \underline{p}$$

where

$$\underline{\psi}_j = (Y_{\rho j\,1} \underline{k}_1{}^T, Y_{\rho j\,2} \underline{k}_2{}^T, \quad . \quad . \quad . \quad , Y_{\rho j\,e} \underline{k}_e{}^T)^T.$$

Hence any point on the surface $f(\alpha,t)$ can be approximated by

$$f^*(\alpha,t) = \underline{F}(\alpha,t) \cdot \underline{p} \qquad (4.2.3)$$

where

$$\underline{F}(\alpha,t) = \underline{\psi}_j(t)\zeta_{0j}(\alpha) + \underline{\psi}_{j+1}(t)\zeta_{1j}(\alpha) + \underline{\varphi}_j(t)\gamma_{0j}(\alpha) + \underline{\varphi}_{j+1}(t)\gamma_{1j}(\alpha) \qquad (4.2.4)$$

for $t_i \leqslant t \leqslant t_{i+1}$ and $\alpha_j \leqslant \alpha \leqslant \alpha_{j+1}$ where x_j is the position of the jth knot as defined in section 3.3 for a natural spline.

The covariance of any two points $f^*(a_\alpha, a_t)$ and $f^*(b_\alpha, b_t)$ can now be obtained from

$$\text{Cov}(a,b) = \underline{F}(a_\alpha, a_\alpha)^T E_r \underline{F}(b_\alpha, b_t), \qquad (4.2.5)$$

where E_r is the diagonal matrix

$$E_r = \text{diag}[\underline{\xi}_1{}^T, \underline{\xi}_2{}^T, \quad . \quad . \quad . \quad , \underline{\xi}_e{}^T].$$

4.3 Characteristics of $f(\alpha,t)$

Thus far we have a means of approximating the surface $f(\alpha,t)$ and its associated uncertainty, but we have yet to make use of the strong correlations which are expected between points on the same "characteristic", i.e. the set of points $\{f(\alpha,\alpha+t_b); \; \alpha > 0, \; t_b \text{ fixed}\}$; the cohort of individuals born at t_b.

A characteristic describes the variation with time of the population in a cohort of individuals born in a short time interval near t_b; since there is no recruitment to a cohort of individuals of ages other than zero, the characteristic $f(\alpha,\alpha+t_b)$ is therefore a monotonically decreasing function of α. Smoothing the surface $f^*(\alpha,t)$ along its characteristics takes advantage of the strong correlation expected between points on such a line. At the same time this smoothing should increase the area of the population surface meeting the monotonicity requirement.

With a little routine algebra, it can be shown from equation (4.1.1) that the gradient of a characteristic at any age and time is the instantaneous total death rate $\mu(\alpha,t)f(\alpha,t)$: thus fitting cubic splines to the characteristics has the effect of minimising

$$\int_{\alpha_0}^{\alpha_e} [\{(\mu(\alpha,\alpha+t_b)f(\alpha,\alpha+t_b))\}']^2 \, d\alpha,$$

among at least the set of C^2 functions sharing the same knot positions. This is roughly equivalent to minimising total rate of change of death rate along a characteristic and provides a way of imposing the condition that death rate should change only slowly with stage.

The most obvious way to fit cubic splines to the characteristics of $f^*(\alpha,t)$ is to calculate a set of points on the characteristic along with the corresponding covariance matrix, Co. A cubic spline can be fitted to these points by replacing the matrix $\mathbf{W}$ by the matix Co^{-1} in equations (3.2.6) to (3.2.9) and choosing ρ by minimisation of (3.3.5). This solution is unsatisfactory for two reasons. Firstly, there are no good criteria for deciding how many points to take: too few and information is lost; too many and matrix computations become expensive and possibly ill-conditioned. Secondly, if we do opt to minimise information loss by fitting the spline to a large number of points on each characteristic, then not only is computation slow but simple algorithms for imposing monotonicity tend to become unworkable.

Having eliminated, or at least argued against, the obvious strategy, we instead use $f^*(\alpha,t)$ to estimate the quantity

$$\upsilon_j(t_b) = \int_{\alpha_{j-1}}^{\alpha_j} f(\alpha,\alpha+t_b) \, d\alpha$$

for each stage. Once again there are many ways of doing this but we choose to find a set of points $\underline{f}_c$ lying along the characteristic to which we fit an interpolating spline. At the same time the covariance matrix $Co(\underline{f}_c)$, of $\underline{f}_c$, can be calculated. By constructing a matrix $\mathbf{H}$ from the integrals of the basis functions of the spline, in the same way as we did in section 3.3, we now obtain

$$\underline{\upsilon} = \mathbf{H}\underline{f}_c \qquad (4.3.1)$$

and

$$Co(\underline{\upsilon}) = \mathbf{H}^T Co(\underline{f}_c)\mathbf{H}. \qquad (4.3.2)$$

A cubic area approximating spline can now be fitted to $\underline{\upsilon}$.

For many datasets the majority of characteristics estimated in this way will decrease monotonically with increasing age, but inevitably some remain recalcitrant. This may be the result of bad stage duration estimates, noise or

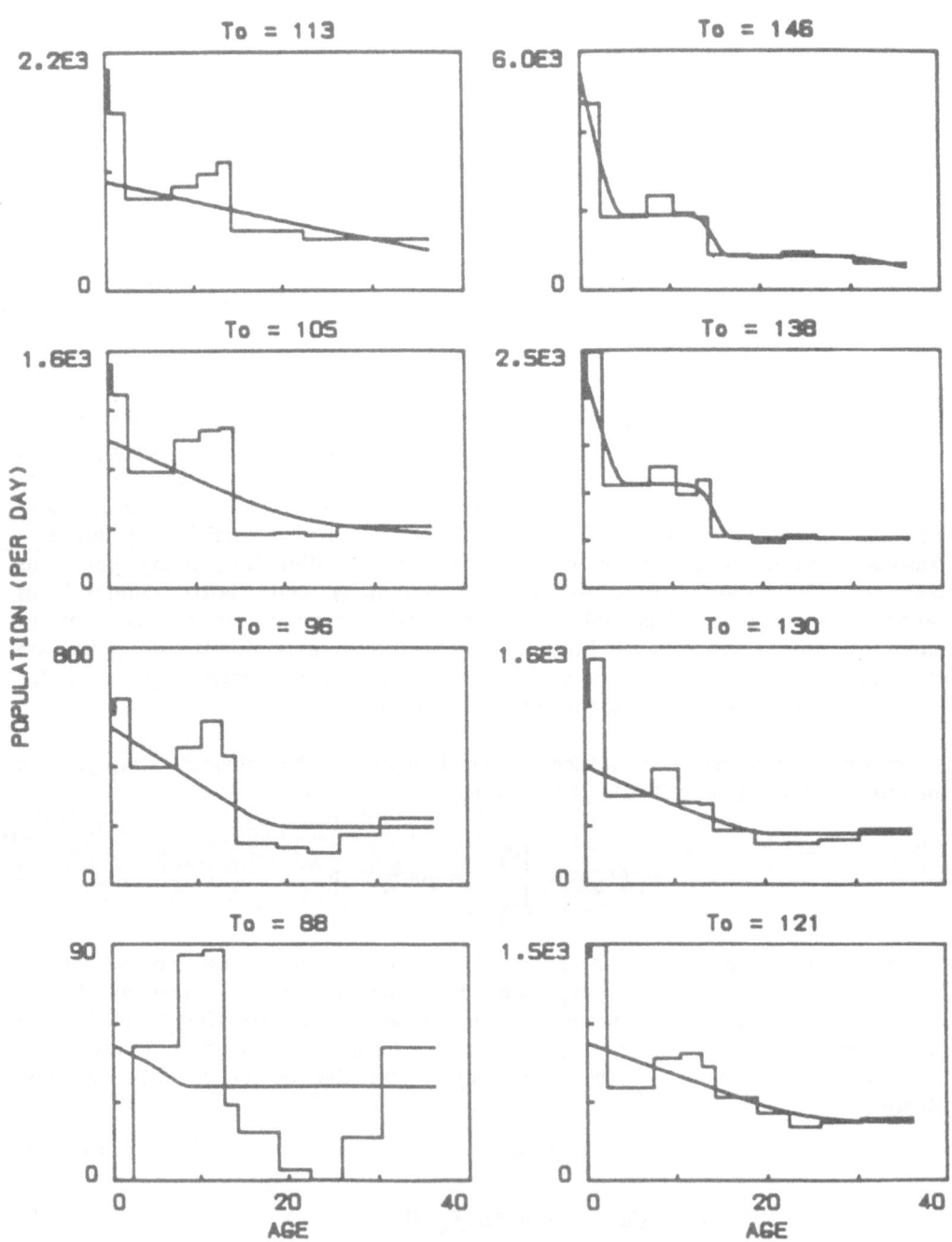

Figure 4.3 Some characteristics from the estimate of $f(\alpha,t)$ for the population used in figures 4.1 and 4.2. The continuous smooth line is the final estimate and for most of the characteristics shown the effect of the monotonicity imposing algorithm of section 4.3 is clear. The histograms show the estimates of integrals along characteristics made using the preliminary estimate of $f(\alpha,t)$. The characteristics start at the $\alpha = 0$, $t = T_0$.

simply data which has corners too sharp for the splines to follow. The ideal solution to the problem would be to find the smoothest monotonic spline function consistent with the data, but whilst such splines exist their actual evaluation is still problematic (Utreras, 1985). We have therefore adopted a fairly crude algorithm for imposing monotonicity on the characteristics:

Initial data consists of knots (x_j, y_j) and weights w_j. Strings of datapoints with the same y_j values are treated as having one weight Σw_j where the summation is over all points in the string. If one point in such a string is moved then all the other values in the string are also changed so that all elements of the string remain equal. Subject to these rules the algorithm is simple:
An index of increase $(y_{j+1} - y_j)/(y_{j+1} + y_j)$ is formed for each interval x_j to x_{j+1} and if any index is greater than zero then the end points of the interval with the highest index are set to the weighted average of the end points $(y_j w_j + y_{j+1} w_{j+1})/(w_j + w_{j+1})$. This process is repeated until the dataset is non-increasing. The weights, w_j, can be set according to the population of the stage in which the knot is located, or the error variance of that stage population.

Once a non-increasing set of points $\underline{y}$ has been obtained, Hyman's (1983) algorithm is used to produce a monotonically constrained, cubic spline based interpolant to these points. Figure 4.3 shows some results from the imposition of monotonicity on the characteristics. In this way we can produce estimates of $f(\alpha,t)$ and $\mu(\alpha,t)f(\alpha,t)$ down any characteristic of f, enabling complete population and total death rate surfaces to be constructed characteristic by characteristic. A pair of such surfaces are shown in figure 4.4. It is also possible to produce an estimate of *the per capita death rate*, $\mu(\alpha,t)$, but this will be undefined when $f(\alpha,t)$ is zero, and may behave unpleasantly as $f(\alpha,t) \to 0$.

4.4 Error estimates

Rough estimates of the variance associated with a characteristic (and hence the surfaces $f(\alpha,t)$ and $f(\alpha,t)\mu(\alpha,t)$) can be obtained using the law of propagation of errors (see, for example, Meyer 1975) in association with the covariance matrix $Co(v)$. Using the matrix Y_ρ (equation 3.3.4) which maps $\underline{v}$ to $\underline{y}$ we obtain a covariance matrix for $\underline{y}$

$$Co(y) = Y_\rho^T Co(v) Y_\rho. \qquad (4.4.1)$$

Defining G_ρ as the matrix mapping $\underline{v}$ to the gradients of the spline approximation for the characteristic we get

$$Co(\mu f) = G_\rho^T Co(v) G_\rho. \qquad (4.4.2)$$

The uncertainty σ_μ in μ can be estimated using a linear approximation. Writing $\sigma_f^2 = Co(y)_{ii}$ and $\sigma_{\mu f}^2 = Co(\mu f)_{ii}$ we obtain

$$\sigma_\mu^2 = \sigma_{\mu f}^2/f^2 + \sigma_f^2 \mu^2/f^2 - 2G_{\rho i i} \sigma_f \mu/f^2,$$

where $G_{\rho i i}\sigma_f$ is the estimated covariance of f and μf at the ith knot of the

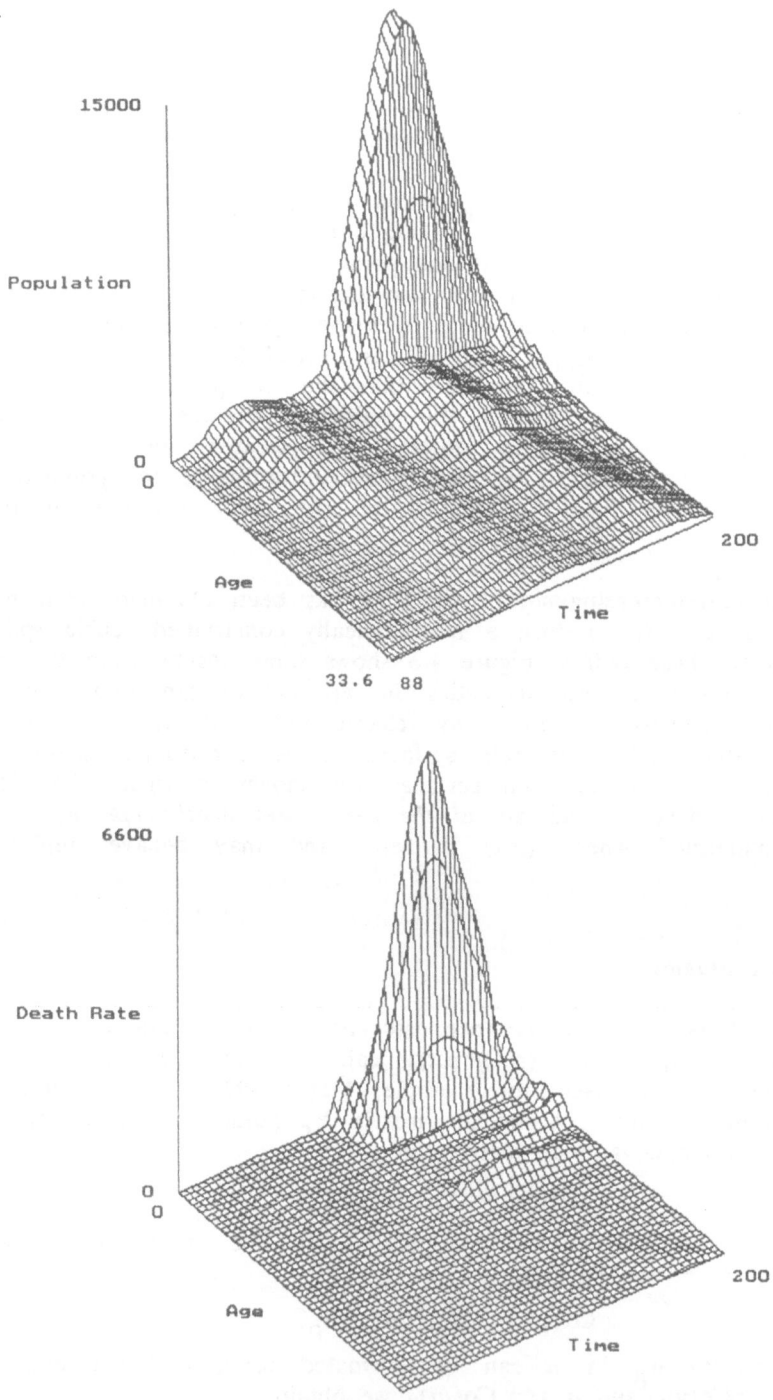

Figure 4.4 The final population and death rate surfaces for the *Pseudocalanus* population used in Figures 4.1 to 4.3.

spline approximating the characteristic.

These estimates do not take into account the effect of the application of the monotonicity algorithm on the error variance, an effect which will vary widely between datasets, depending on how frequently the algorithm has to be applied for a given dataset. Furthermore the variance estimates only measure the sensitivity of our estimates to random measurement error. Such estimates say nothing about the errors which arise from assuming the wrong model for the population.

4.5 What to do about adults

The adult stage of most species is not of fixed duration: for example, with copepods most individuals leave adulthood when they get eaten. For this reason the population surface mortality estimation method is inappropriate for this final stage. What the population surface does yield is an estimate of recruitment to the adult stage, from the surface boundary at α_{e+1}. This time dependent recruitment estimate can be fed into a scheme of the type described by equation (2.2.2) to obtain adult mortality rates.

4.6 A simpler method

The method as outlined in sections 4.2 and 4.3 is fairly complicated and in some circumstances a considerable simplification may prove almost as effective and very much faster. If we assume that locally $f(\alpha,t)$ is not too far from being planar then we can write

$$v_j(t_b) = \eta_j \{t_b + (\alpha_j + \alpha_{j+1})/2\}$$

and assume that the error variance of $v_j(t_b)$ is equal to the variance of $\eta_j\{t_b + (\alpha_j + \alpha_{j+1})/2\}$ and is independent of the error variance of other stages. In chapter 5 we compare this simplification to the full method with numerical tests.

4.7 Messy practicalities

There are several practicalities which further complicate the surface fitting method: small numbers of datapoints for each stage, unreliable ρ estimates for the characteristics, whether to use area–approximating splines or area–matching splines to reconstruct age structures, and what to do at the corners of the surface.

Stage time series

Ideally, splines would be fitted to each stage independently, but in practice the small number of datapoints for each stage precludes this. Wahba's 1983 study indicated that cross validation becomes increasingly unreliable when there are below about thirty datapoints: the technique tends to over estimate the smoothing parameter, ρ, so that the data are under-smoothed.

The reasons for this are fairly simple. Over any small number of points there is a chance that the noise affecting a dataset will appear correlated. Since this effect is local it will not seriously bias ρ estimates for datasets with many points, but if there are few points in the dataset the data will appear less noisy than it really is and the ρ estimate will be too high. A moment's consideration shows that the "opposite" circumstance of "perfectly uncorrelated" noise has no detrimental effect on our estimates (it is in fact exactly what we want). For few datapoints we therefore expect the distribution of ρ estimates to be positively skewed with respect to ρ. This skewing is made worse by the shape of the function $G_{cv}(\rho)$ (defined in equation (3.2.12). Figure 4.5 shows $G_{cv}(\rho)$ against ρ for the twelve stages of a population of the copepods *Pseudocalanus* (Loch Ewe 1980 bag C1 see Hay *et. al.* 1988). For stages with high ρ estimates the minimum of $G_{cv}(\rho)$ is highly asymmetric, being very shallow towards high ρ values. Given that $G_{cv}(\rho)$ is itself calculated from a few noisy datapoints this must mean that there will be a further increase in the skew of the distribution of ρ estimates.

The reliability of an estimate of ρ can be increased if there are grounds for assuming that ρ will take approximately the same value for each stage. This is roughly equivalent to a constant ratio of mean square population to variance across the stages (see equation (3.2.6)). Whilst such an assumption would be erroneous for a dataset subject only to Poisson sampling error, it does conform to expectation for a population distributed between a few patches, where the number of patches rather than the number of animals is likely to make the chief contribution to variance.

Assuming then that ρ is the same for each stage but that the estimate of ρ is biased upwards of the true value, we estimate ρ for each stage by cross validation and then choose the lowest value found, using this to fit the splines which will be used for stage population and error variance estimation. Alternatively, if we really do want to assume a Poisson type error model then we can force the ratio of ρ to the mean stage population to be the same for each stage.

What type of spline for age structure reconstruction?

In section 4.2 we left open the question of whether area-matching or area-approximating splines are more appropriate for the preliminary estimation of age structure. Given noisy estimates of the age structure histograms at any time, some form of smoothing would probably improve an estimate of continuous age structure, but if we fit smoothing splines to the data by minimisation of (3.2.11) then we are effectively seeking the smoothest age distribution which can be fitted to the data consistent with its estimated error variance. There is no reason to suppose that this is appropriate: our initial assumption was that death rate would vary slowly between stages, a condition which does not require smooth age structures. Given this difficulty in choosing

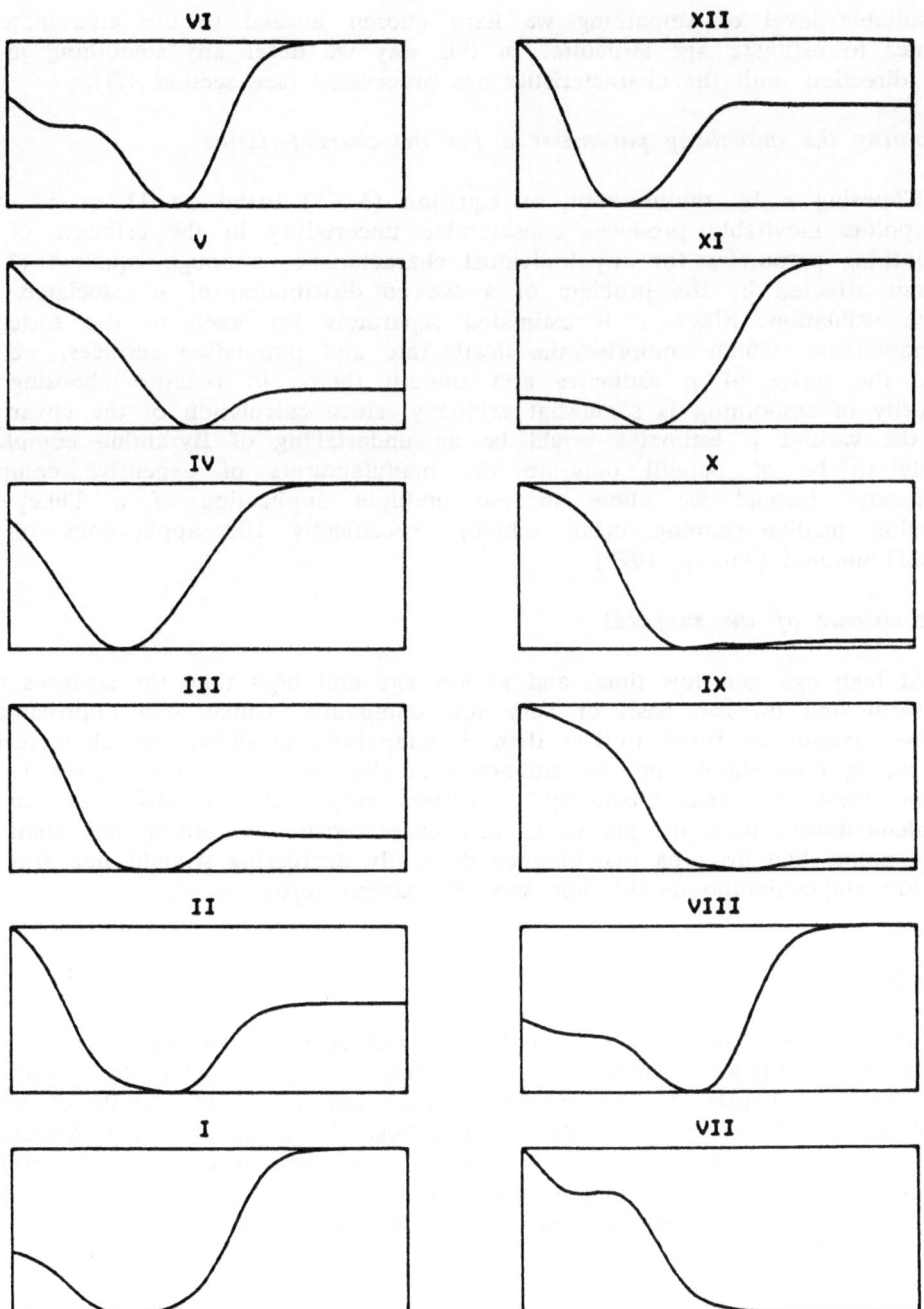

Figure 4.5 The shape of the function $G_{cv}(\rho)$ against ρ for the twelve stages of the copepod population shown in figure 4.1. The horizontal axis is $\log(\rho)$, the scale being the same for each stage.

a suitable level of smoothing we have chosen instead to use area–matching splines to estimate age structure. In this way we defer any smoothing in the age direction until the characteristics are proccessed (see section 4.3).

Choosing the smoothing parameter ρ for the characteristics

Choosing ρ by minimisation of equation (3.3.5) based on 11 or 12 noisy datapoints inevitably produces considerable uncertainty in the estimate of the smoothing parameter for any individual characteristic, although equation (3.3.5) is not affected by the problem of a skewed distribution of ρ associated with cross validation. Since ρ is estimated separately for each of the series of characteristics which comprise the death rate and population surfaces, we can take the series of ρ estimates and smooth them. In practice choosing the severity of smoothing is somewhat arbitrary, since calculation of the covariance of the various ρ estimates would be an undertaking of Byzantine complexity likely to be of benefit only to the manufacturers of expensive computer hardware. Instead we chose to use multiple application of a Tukey–type running median–running mean scheme: specifically 100 applications of the 4253H method (Tukey, 1977).

The corners of the surfaces

At high age and low time, and at low age and high time the surfaces must be estimated on the basis of very few datapoints. Cubic area approximating splines cannot be fitted to less than 3 datapoints, so either the characteristics of the surface should not be smoothed at the corners or alternatives to the spline must be used. Obviously it matters very little, in any case. In the implementation used in this work, any characteristic comprising less than two data points had an area matching conditionally decreasing straight line fitted to it. For single datapoints this line was of gradient zero.

4.8 Summary

We have presented in gory detail a method of reconstructing population and death rate surfaces, which uses spline functions to combat the instabilities identified in chapter 2. The method is only one of a rich family of similar schemes and is almost certainly not the best. What is important however is how well the method performs in practice, and how it compares to previous approaches. This assessment and comparison is the subject of the next chapter. We end this chapter with a formal outline of the method.

Recipe for death rate estimation

INPUT: Time series of stage populations, $\underline{p}_j$, and estimates of stage durations.

STEP 1:   Fit a cubic smoothing spline to each stage population time series by the method of generalized cross validation (equation 3.2.12) to obtain estimates of the smoothing parameter, ρ, for each stage.

STEP 2:   Choose the lowest (or second lowest) value of ρ and use this to refit smoothing splines to all the stage time series to find $\mathbf{A}_\rho$ (3.2.9), $\mathbf{C}_\rho$ (3.2.10) and $\hat{\sigma}^2$ (3.2.13) for each stage.

STEP 3:   Form the vectors of population and estimated sample variance defined by (4.2.2) and (4.2.1) respectively.

STEP 4:   Calculate the matrices $\mathbf{Y}_\rho$ and $\mathbf{M}_\rho$ which map stage population data at a given time to spline coefficients for continuous age structure estimates.

STEP 5:   Repeat steps 7 to 9 for enough characteristics to form estimates of $f(\alpha,t)$, $f(\alpha,t)\mu(\alpha,t)$ and $\mu(\alpha,t)$, storing the smoothing parameters ρ for each estimate.

STEP 6:   Smooth the stored values of ρ by some Tukey type time–series method. Repeat steps 7 to 11 for each characteristic but this time using the stored smooth values of ρ rather than finding ρ.

STEP 7:   Use equations (4.2.3) and (4.2.4) with the population vector to find a series of points, $\underline{f}_c$, along a characteristic (for example one point at each stage boundary and one in each stage midpoint). Use equation (4.2.5) with the vector of variance estimates to find the asscociated covariance matrix $Co(f_c)$.

STEP 8:   Estimate the integral of the characteristic across each stage, v_j, using $\underline{f}_c$ and the matrix $\mathbf{H}$ (equation 4.3.1). $\mathbf{H}$ is defined in the same way as Λ in section 3.3.1. Use equation (4.3.2) to estimate the covariance marix of $\underline{v}$, $Co(\nu)$.

STEP 9:   Fit an area–approximating spline (section 3.3.2) to $\underline{v}$ choosing ρ by substituting $Co(\nu)$ into (3.4.1) and minimising with respect to ρ.

STEP 10: Estimate the uncertainties associated with the resulting population and death rate estimates as described in section 4.4.

STEP 11: If the characteristic estimate is anywhere non–decreasing with age then apply the monotonicity imposing algorithm of section 4.3 to it.

STEP 12: Obtain stage rates by numerical integration of the population and death rate surfaces.

OUTPUT: Surfaces describing the temporal development of population age structure and the time and age dependence of death rate and, if required, *per capita* death rate; error estimates for these surfaces; time series of estimated stage specific population and death rates.

CHAPTER 5

TESTS OF THE NEW METHOD

In chapter 4 a new mortality rate estimation scheme was developed. This new method will now be compared to methods proposed by Parslow, Sonntag and Matthews (1979), Manly (1987) and Hay, Evans and Gamble (1988) using simulated data which for most of the comparisons, we constructed in a way that would, *a priori* favour the "rival" method. The two variants of the new method of chapter 4 will then be compared in a similar way.

5.1 Parslow and Sonntag's Lag–Manly model method

Parslow, Sonntag and Matthews (1979) proposed estimating mortality and stage duration for a population by fitting a model to noisy stage structured census data using the approach known as 'systems identification' (see section 1.4). They tried several models, but the most successful appeared to be their 'Lag–Manly' model. This is a simple stage structure model of the type described by equations (2.1.1) to (2.1.4). Recruitment to the first stage by is represented by a Gaussian:

$$R_1(t) = \frac{\Omega_1}{\sqrt{2\pi}\ \sigma_1} \exp\{-(t-t_0)^2/2\sigma_1{}^2\}. \qquad (5.1.1)$$

The *per capita* death rate for each stage is represented by a stage–dependent, time–invariant parameter, μ_i, where i is the stage index. For our reconstruction, the parameters in (5.1.1) along with the stage durations τ_i and the parameters μ_i were estimated using Marquardt's method (see for example, Press *et al*. 1986). This method minimises chi–squared for the model fit to data by combining a steepest descent method with a method based on explicit minimisation of a quadratic form approximating the dependence of chi–squared on the model parameters. The method requires estimates of the partial derivatives of the model solution with respect to the parameters at each of the datapoints. For most of the parameters this involves solving the differential equations resulting from partial differentiation of the appropriate stage population equation (2.1.1) with respect to the model parameter. The differentials with respect to some of the mortality parameters are more easily calculated. Using some results from Gurney, Nisbet and Lawton (1983) it is easy to show that

$$\frac{d\eta_j}{d\mu_i} = -\tau_i\eta_j(t) \quad \text{for} \quad i<j$$

$$\text{and} \quad \frac{d\eta_j}{d\mu_i} = 0 \qquad \text{for} \qquad i>j.$$

We used the implementation of Marquardt's method given in Press *et al*. (1986), with the obvious and slight modification that the model and its derivatives were evaluated in a single call per parameter set rather than a call for each datapoint. (Note also an error in Press *et al*.'s routine: the line which reads 'ATRY(LISTA(J)) = ATRY(LISTA(J))+DA(J)' should read 'ATRY(LISTA(J)) = A(LISTA(J))+DA(J)' if you would prefer the method to converge. This error is corrected in the later editions of Press *et al*.) The chi–squared statistic minimised by the Marquardt method was set up so that the contribution from each stage was weighted by the reciprocal of the average population for the stage. This weighting was intended to avoid domination of the fit by the early juvenile stages at the expense of later stages. The stage durations, τ_j, were treated as fixed known parameters so that the comparison with the method of chapter 4 would be fair. All differential equations in the model were integrated numerically using the fourth order Runge–Kutta method.

Comparison of the new method with Parslow et al.'s method

To facilitate comparison of Parslow *et al*.'s method with the new method, artificial data were generated. Three sets of parameters were chosen for their "Lag–Manly" model reflecting roughly the expected pattern of mortality in a copepod dataset whose dynamics could be described by such a model. The parameters are given in table 5.1: stage durations were chosen to reflect the range of estimates given for the copepod *Pseudocalanus* in the literature; death rates in set 1 decline fairly smoothly with age; in set 2 this smoothness is violated by stages 2 and 6; in set 3 the death rates violate the smoothness assumptions of the new method by varying erratically between stages. The three models were solved numerically using the fourth order Runge–Kutta method (see, for example, Burden and Faires, 1985) and both stage populations and stage specific total death rate were calculated. Adult stages were not considered since the method of chapter 4 deals only with pre–adult stages and uses the method of chapter 2 for the adult stage, tests of this latter method are given in chapter 2.

Each artificial dataset was peturbed by gaussian noise, generated using routines given in Press *et al*. (1986). Two noise levels were used: the standard deviation for each stage was set to 20% and 40% of the mean stage population for that stage. Five replicates of each combination of noise level and parameter set were produced, yielding 30 datasets in total.

Stage	Set 1 μ	Set 1 τ	Set 2 μ	Set 2 τ	Set 3 μ	Set 3 τ
σ_1	15.0		10.0		17.0	
Ω_1	500,000		400,000		500,000	
t_0	140.0		130.0		145.0	
1	0.3	0.8	0.1	1.2	0.05	2.0
2	0.25	1.2	0.2	1.2	0.3	2.0
3	0.3	3.6	0.1	2.0	0.05	2.0
4	0.1	3.2	0.05	3.2	0.1	3.2
5	0.1	3.2	0.05	3.2	0.1	3.2
6	0.05	2.0	0.15	2.0	0.05	3.2
7	0.05	4.0	0.05	4.8	0.03	3.6
8	0.05	3.2	0.05	3.6	0.1	3.6
9	0.04	3.2	0.05	3.2	0.02	3.6
10	0.03	3.2	0.1	3.6	0.05	3.6
11	0.03	3.6	0.1	4.8	0.04	3.6

Table 5.1 The 3 parameter sets used in test Lag-Manly data.

The new method and Parslow *et al.*'s method were applied to each dataset to estimate stage specific time dependent total death rate. For the sake of speed the simplified version of the new method (section 4.5) was used. Three statistics were calculated:

$$RMS = \sum_{j=1}^{e} \frac{\{\Sigma(\mu_{ji}-\mu_{ji}^{est})^2/n\}^{\frac{1}{2}}}{\Sigma\mu_{ji}/n} ,$$

$$MeanABS = \sum_{j=1}^{e} \frac{\Sigma|\mu_{ji}-\mu_{ji}^{est}|}{\Sigma\mu_{ji}} \quad \text{and}$$

$$\text{MaxDev} = \text{MAX} \left\{ \frac{|\mu_{ji} - \mu_{ji}^{est}|}{\Sigma \mu_{ji}/n} \right\},$$

where μ_{ji} is the death rate in the jth stage at the ith sample time.

Table 5.2 summarises the results of these tests, expressing the statistics for Parslow *et al.*'s method as multiples of the equivalent statistic for the new method averaged over the replicates. In the 90 comparisons which contributed to table 5.2 the Parslow *et al.* method did better than the new method in only 4 cases. For the sake of completeness it was also confirmed that the Parslow's method could reconstruct all the parameters in table 5.1 in the absence of noise.

Noise level	set	RMS	MeanABS	MaxDev
20%	1	3.4	3.2	3.0
	2	3.5	3.3	3.3
	3	1.7	1.8	1.5
40%	1	2.9	2.8	2.6
	2	3.6	3.3	3.3
	3	1.9	2.0	1.5
Mean		2.8	2.7	2.5

Table 5.2 The errors in parameter estimates from Parslow
et al.'s method as multiples of the equivalent
error using the new method.

The results reported in table 5.2 are from an implementation of Parslow *et al.*'s method which does not constrain the death rate estimates to be positive. This constraint was introduced in a second set of tests which also used the parameters given in table 5.1. The model was fitted to the data and if any death rate estimate was negative then the model was re-fitted to the data with the most negative death rate fixed at zero. This proccess was repeated until none of the death rate estimates were negative. Table 5.3 is the equivalent to 5.2 for this implementation of the method. Out of 90 comparisons the constrained Parslow *et al.* method was an improvement on the new method in 19. 10 of these cases were in the comparison of maximum deviations, probably because of the slightly erratic behaviour of the new method at the the corners of the death rate surface, where estimates are made on the basis of very few datapoints.

58

Noise level	set	RMS	MeanABS	MaxDev
20%	1	1.3	1.3	1.1
	2	1.9	1.9	1.8
	3	1.2	1.2	1.01
40%	1	1.5	1.5	1.1
	2	1.7	1.6	1.6
	3	1.1	1.2	0.97
Mean		1.5	1.5	1.3

Table 5.3 The errors in parameter estimates from Parslow
et al.'s method as multiples of the equivalent
error using the new method.

Conclusions

The model from which the simulation data for these tests was generated was exactly the model which is fitted to the data by the method of Parslow *et. al*. Even the integration method was identical in simulation and fit. Despite this seemingly ideal test data the Parslow *et. al*. method performed poorly in comparison to the new method. Examination of the actual output parameters suggests that this can probably be attributed to the age-propagating instabilities identified in chapter 2. Figure 5.1 shows a typical set of stage specific total death rate estimates from Parslow *et. al*.'s method plotted with the "true" total death rates shown for comparison. The reconstruction is from data simulated using parameter set 2 from table 5.1 which was perturbed by gaussian noise with standard deviation for each stage set to 20% of the stage mean. Note that for three of the stages (I, VIII and X) the death rate had to be constrained to zero to prevent negative death rate estimates. The estimates display the 'undershoot-overshoot' characteristic of age-propagating instability. Furthermore the errors in the estimates show a tendency to increase in magnitude with stage, as expected if age propagating instability is at work, although the effect is moderated by the constraint that no estimate shall be less than zero.

Given the clear evidence that Parslow *et. al*.'s method is unlikely to produce good death rate estimates we turn instead to Hay *et. al*.'s (1988) modified version of Parslow *et. al*.'s method.

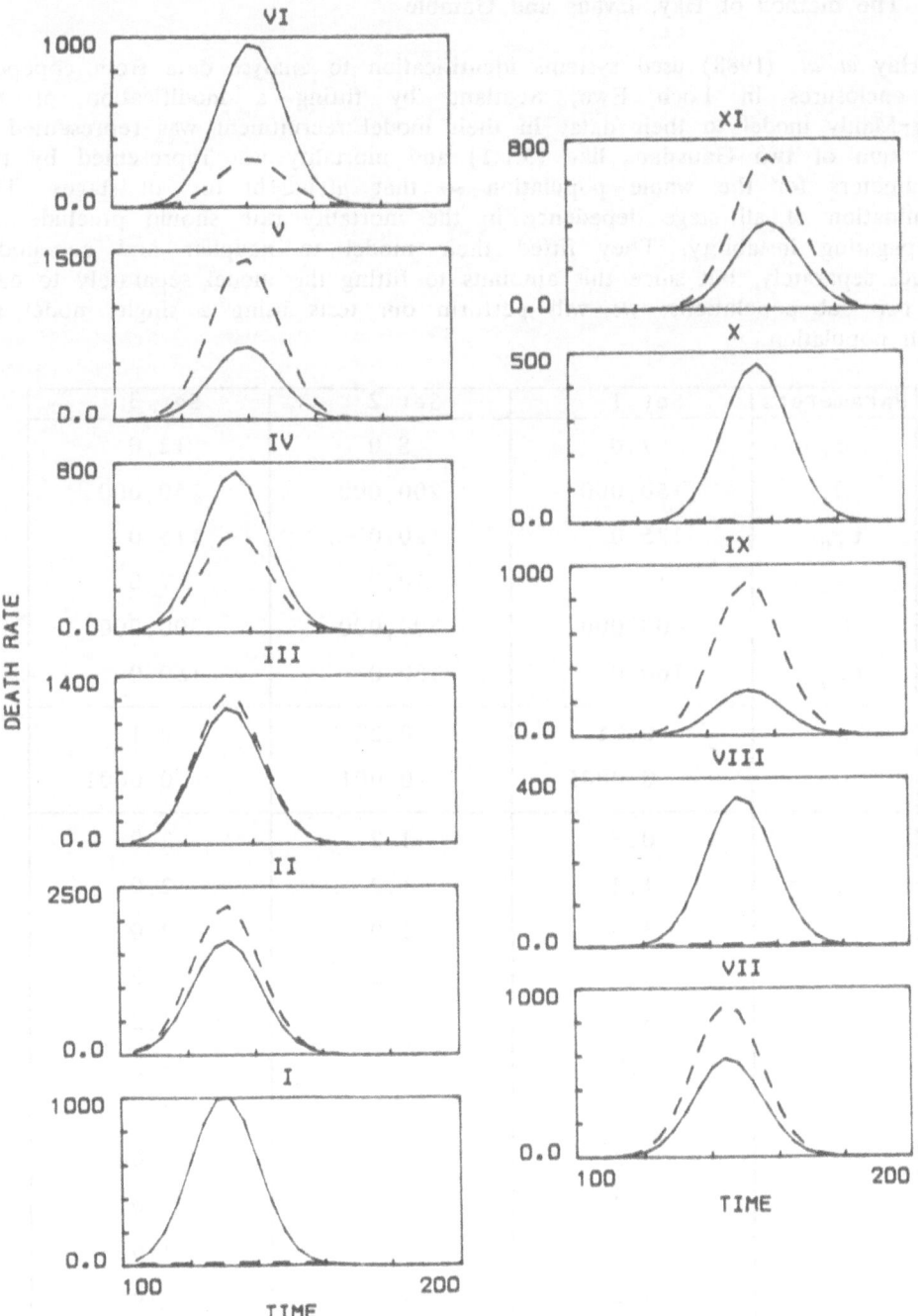

Figure 5.1 An example of the death rate estimates produced by fitting the Lag—Manly model of Parslow *et.al.* to simulation data, clearly showing age-propagating instability. The death rate estimates are constrained to be non-negative. The simulated population data for each stage was perturbed by gaussian noise of standard deviation equal to 20% of the mean stage population. The continuous line is the true death rate and the dashed line the estimated death rate.

5.2 The method of Hay, Evans and Gamble

Hay *et al.* (1988) used systems identification to analyse data from copepods in enclosures in Loch Ewe, Scotland by fitting a modification of the Lag-Manly model to their data. In their model recruitment was represented as the sum of two Gaussians like (5.1.1) and mortality was represented by two parameters for the whole population so that $\mu(t)=a+bt$ for all stages. This elimination of all stage depedence in the mortality rate should preclude age propagating instability. They fitted their model to naupliar and copepodite stages separately, but since this amounts to fitting the model separately to each of two sub-populations, we will perform our tests using a single model for each population.

Parameters	Set 1	Set 2	Set 3
σ_1	7.0	8.0	12.0
Ω_1	150,000	200,000	150,000
t_{10}	125.0	120.0	115.0
σ_2	14.0	10.0	7.0
Ω_2	400,000	300,000	300,000
t_{20}	160.0	160.0	150.0
a	0.01	0.22	0.1
b	0.0005	-0.001	0.0001
τ_1	0.8	1.2	2.0
τ_2	1.2	1.2	2.0
τ_3	3.6	2.0	2.0
τ_4	3.2	3.2	3.2
τ_5	3.2	3.2	3.2
τ_6	2.0	2.0	3.2
τ_7	4.0	4.8	3.6
τ_8	3.2	3.6	3.6
τ_9	3.2	3.2	3.6
τ_{10}	3.2	3.6	3.6
τ_{11}	3.6	4.8	3.6

Table 5.4 The 3 parameter sets used in test Hay-Evans data. The first six parameters (σ_1 to t_{20}) describe the recruitment (cf. equ. 5.1.1), the parameters a and b characterise mortality as described in the text, and the parameters τ_1 to τ_{11} are stage durations.

We compared the method to the new method in exactly the same way as we compared the method of Parslow *et. al.* to the new method. That is, the model to be fitted to the data was again used to produce simulation data, which was perturbed by random error. Table 5.4 gives the parameter sets used for the simulated data. Table 5.5 summarises the results of comparing the two methods by giving the error statistics for Hay *et. al.*'s method as multiples of the statistics for the new method.

Noise level	set	RMS	MeanABS	MaxDev
20%	1	0.09	0.09	0.07
	2	0.05	0.06	0.04
	3	0.14	0.15	0.13
40%	1	0.19	0.2	0.15
	2	0.1	0.8	0.8
	3	0.2	0.2	0.19
Mean		0.12	0.13	0.11

Table 5.5 The errors in parameter estimates from Hay
et. al.'s method as multiples of the equivalent
error using the new method.

Clearly the method of Hay *et. al.* does considerably better than the new method. In fact it was better than the new method for all 90 comparative statistics. But these tests are only half the story: for any real dataset we cannot know that the mortality rate is a linear function of time, or that recruitment is the sum of two Gaussians. There is a wide family of models which could have produced fairly similar datasets. What happens if we choose models of slightly different structure which produce datasets of roughly similar appearance? We examined this question by producing three new datasets whose recruitment functions were:

$$R_1(t) = \frac{3000}{1+0.01(t-125)^2} + \frac{6000}{1+0.007(t-160)^2} + 500 + 2e^{0.03t},$$

$$R_2(t) = \frac{2000}{1+(10e^{-\{0.075t-5.17\}})^2} + \frac{6000}{1+(100e^{-\{0.05t-1.5\}})^2} + 300 + e^{0.03t}$$

and $R_3(t) = R_3^*(t) + 500 + 5e^{0.03t}$ where

$$R_3^*(t) = \begin{cases} 0 & t<100 \\ 2000\{1-\cos[\pi t/20]\} & 100 \leqslant t<140 \\ 4000\{1-\cos[\pi t/20]\} & 140 \leqslant t<180 \\ 0 & 180 \leqslant t. \end{cases}$$

The mortality rates were also modified by the addition of a stage specific constant to the death rate term assumed in the Hay *et al.* model. These stage dependent mortality parameters are given in table 5.6 along with the stage durations for each dataset.

Stage	Set 1		Set 2		Set 3	
	μ	τ	μ	τ	μ	τ
1	0.3	1.2	0.3	2.0	0.4	1.2
2	0.2	2.0	0.25	2.0	0.35	2.0
3	0.2	3.2	0.25	3.2	0.2	3.6
4	0.2	3.2	0.2	3.2	0.1	3.2
5	0.05	3.2	0.15	3.2	0.1	3.2
6	0.2	2.4	0.14	3.2	0.05	2.4
7	0.05	3.6	0.1	3.6	0.1	3.6
8	0.05	3.6	0.05	3.6	0.05	3.2
9	0.03	3.6	0.03	3.6	0.03	3.2
10	0.03	4.0	0.03	3.6	0.03	3.2
11	0.05	4.0	0.05	3.6	0.02	4.0

Table 5.6 The 3 parameter sets used in the second set
of tests of the Hay *et al.*'s method.

Figures 5.2 to 5.4 show the simulated datasets produced by these models and table 5.7 presents the results of comparison of Hay *et al.*'s method to the new method using this data. For the RMS statistic the new method improved on Hay *et al.*'s method in 23/30 cases. For AbsDev and MaxDev the equivalent figures were 24/30 and 16/30 respectively.

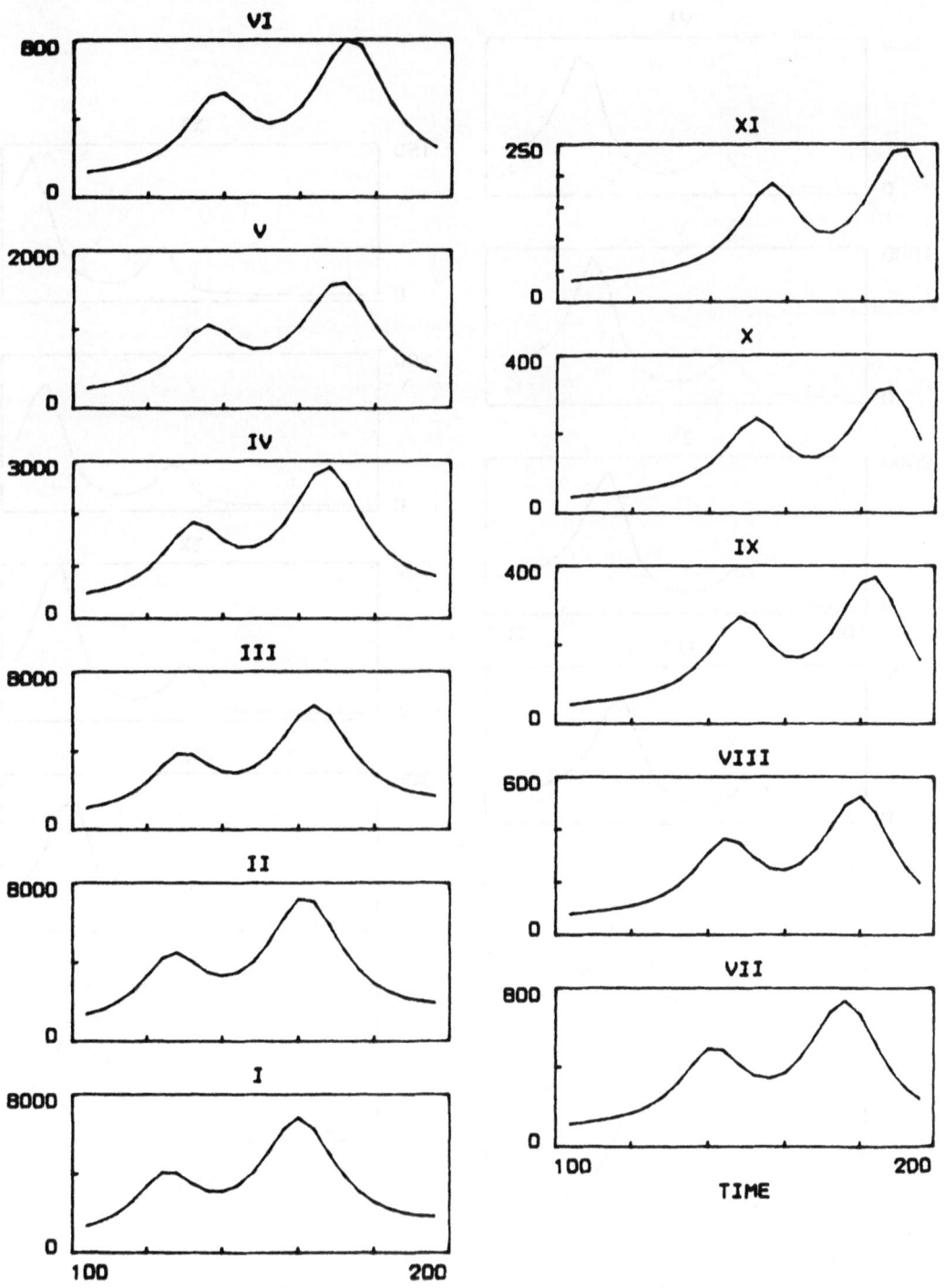

Figure 5.2 Set 1 of the simulated data used in the second batch of tests of Hay *et. al.*'s method. Details are given in section 5.2 and table 5.6.

64

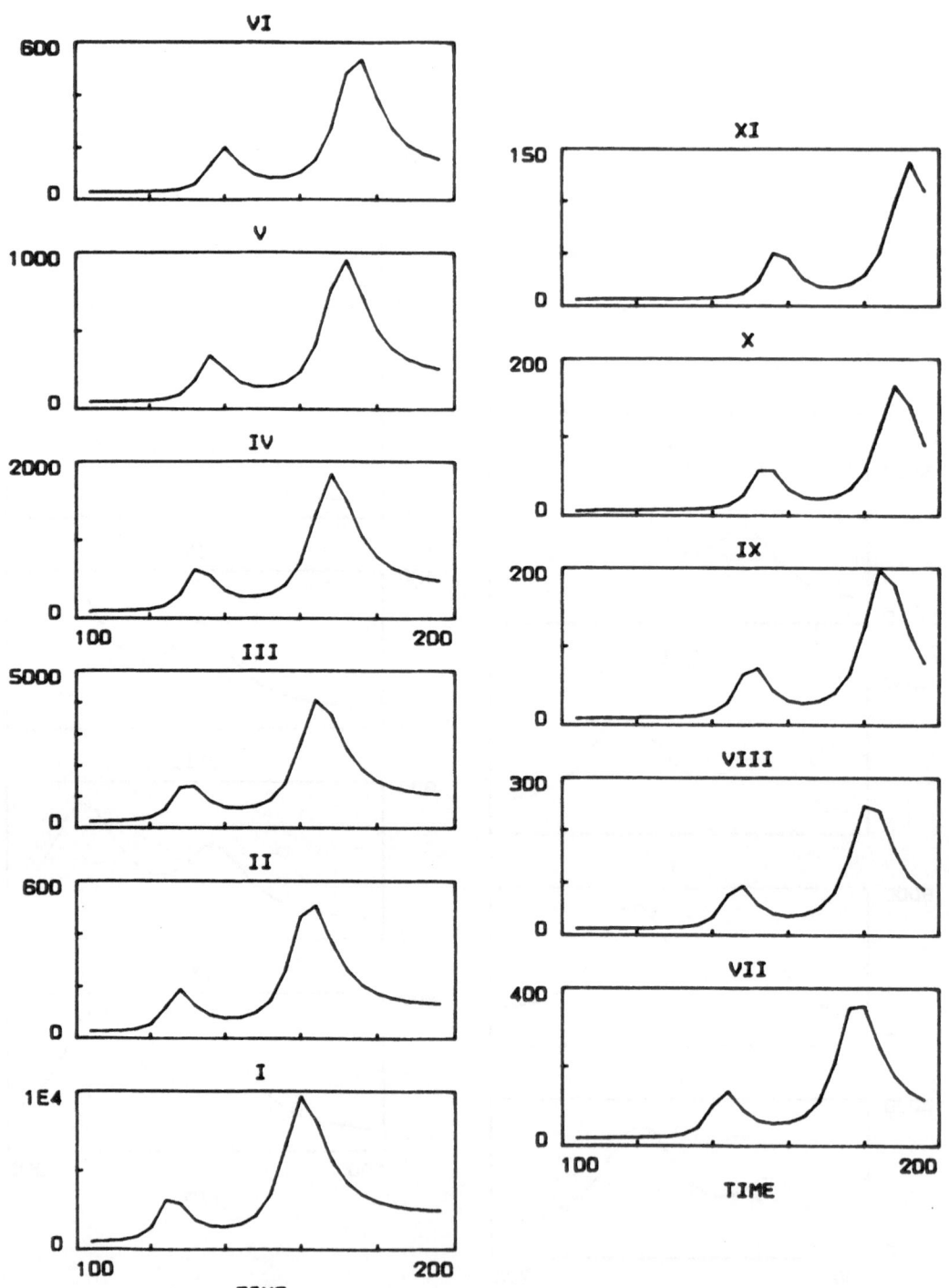

Figure 5.3 Set 2 of the simulated data used in the second batch of tests of Hay *et. al.*'s method. Details are given in section 5.2 and table 5.6.

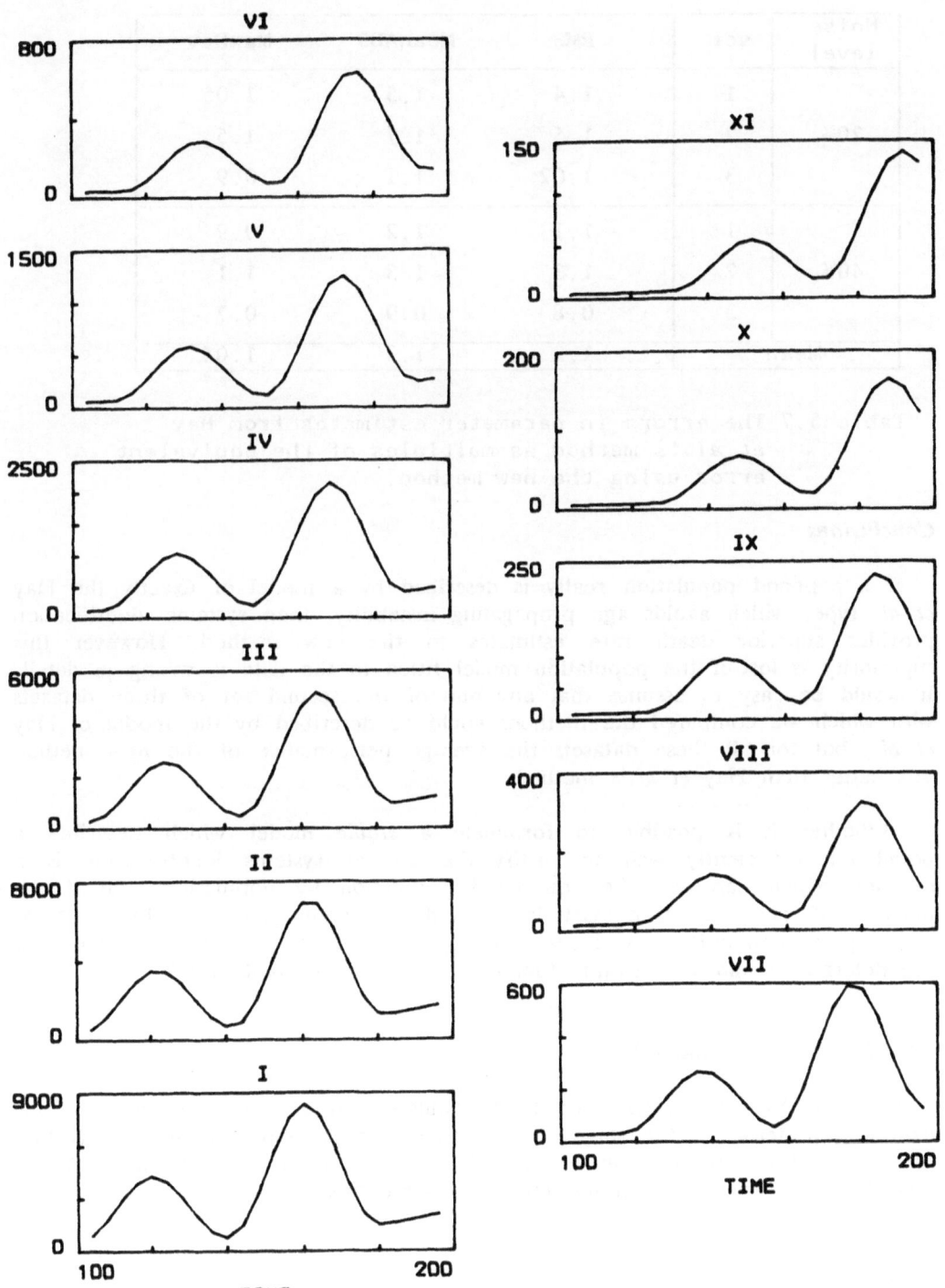

Figure 5.4 Set 3 of the simulated data used in the second batch of tests of Hay *et. al.*'s method. Details are given in section 5.2 and table 5.6.

Noise level	set	RMS	MeanABS	MaxDev
20%	1	1.4	1.5	1.05
	2	1.5	1.7	1.5
	3	1.02	1.1	0.9
40%	1	1.1	1.2	0.9
	2	1.3	1.3	1.1
	3	0.8	0.9	0.7
Mean		1.2	1.3	1.04

Table 5.7 The errors in parameter estimates from Hay *et al.*'s method as multiples of the equivalent error using the new method.

Conclusions

If a copepod population really is described by a model of exactly the Hay *et al.* type, which avoids age propagating instability, then systems identification provides superior death rate estimates to the new method. However this superiority is lost if the population model fitted to the data is wrong in detail. It would be easy to assume that any one of the second set of three datasets with which we compared the methods could be described by the model of Hay *et al.*, but for all these datasets the average performance of the new method was better than Hay *et al.*'s method.

Whether it is possible to formulate a *stable* model which describes a population sufficiently well to justify the use of systems identification is a question which can only be considered population by population, but if the purpose of analysing a dataset is to find out about the mortality patterns affecting a population, is it sensible to perform the analysis assuming a pre-determined and very simple functional form for mortality rate?

5.3 Manly's (1987) method

Manly (1987) argued that the total population at sample time $i\Delta t$ could be expressed in terms of the stage populations at the previous sample time $(i-1)\Delta t$ using the stage specific between-sample survivorship parameters S_j and the recruitment between the sample times. He wrote this as

$$\sum_{j=1}^{e} \eta_{ji} = R_{i-1} + S_1 \eta_{1i-1} + S_2 \eta_{2i-1} + \cdot \cdot \cdot \cdot \cdot + S_e \eta_{ei-1} ,$$

in which S_j is the proportion of those individuals present in the jth stage at one sample time that survive to the next sample time. R_{i-1} is the number of individuals recruited to the first stage between time $(i-1)\Delta t$ and $i\Delta t$. Manly proposed that this quantity should be a piecewise linear function of time so that any value R_{i-1} would be a known linear combination of at most two parameters. Provided that there are fewer recruitment and survivorship parameters to estimate than datapoints then these parameters can be estimated by linear regression since the η_{ji}'s are known.

Simulation data

Data were simulated for comparative tests of Manly's method with the new method in such a way that the assumptions of Manly's method were met exactly. Data were simulated from time 100 to 200 and sampled every 4 time units. For convenience, all stage durations were also set to 4. The recruitment parameter was set at times 100, 120, 140, 160, 180 and 200 and was obtained by linear interpolation between these times. Given this set-up simulation becomes a mere book keeping exercise. Manly suggested that his method was most successful working with stages which were lumped together and assumed to have the same survivorship. For this reason we simulated 12 stages but divided them into 4 groups of three stages having the same survivorship. The data was perturbed by noise in the same way as in the previous comparative tests. The parameter sets used are given in table 5.8.

Parameters	Set 1	Set 2	Set 3
R_1	0.0	0.0	0.0
R_2	500.0	1500.0	1500.0
R_3	1500.0	500.0	500.0
R_4	3000.0	3000.0	3000.0
R_5	2500.0	2000.0	2000.0
R_6	0.0	0.0	2000.0
S_{1-3}	0.8	0.5	0.6
S_{4-6}	0.85	0.8	0.7
S_{7-9}	0.9	0.7	0.9
S_{10-12}	0.2	0.7	0.7

Table 5.8 The parameter sets used to simulate Manly test data

Results

The 30 test datasets were analysed using Manly's method and the new method to obtain estimates of S_{1-3}, S_{4-6}, S_{7-9} and S_{10-12}. Table 5.9

records the average of the absolute error in estimates produced by Manly's method as multiples of the equivalent estimate produced by the new method. In 9 cases out of 120 comparisons the new method performed more poorly than Manly's method. Tests were also performed with noise free data, which confirmed that the Manly method could reconstruct parameters exactly in ideal conditions.

Noise level	set	Ratio of Error in estimates of:			
		S_{1-3}	S_{4-6}	S_{7-9}	S_{10-12}
	1	9.5	7.2	1.8	22
20%	2	385	3.7	4.8	3.0
	3	9.3	5.1	3.6	7.6
	1	895	8.8	6.1	18
40%	2	11	2.6	21	3.2
	3	20	5.3	2.5	7.7
Mean		220	5.4	6.7	10

Table 5.9 The errors in parameter estimates from Manly's method as multiples of the equivalent error using the new method.

Conclusions

Manly's method doesn't really seem to be a contender. A variant of age propagating instability could be to blame since overestimation of S for one group of stages is easily offset by underestimation of the parameter for the next stage.

5.4 Comparison and testing of the sophisticated and simplified versions of the new method

Having demonstrated the merits of the simplified version of the new method relative to some previous methods for estimating stage specific death rates, we now wish to test the ability of both versions of the method to reconstruct death rate and population surfaces and to estimate the standard deviations associated with these surfaces. We also compare how effectively the two versions of the method reconstruct stage specific death rate.

Simulation of data

The data on which the methods were tested were produced by solving the von Foerster equation (4.1.1) for three different *per capita* death rate functions $\mu(\alpha,t)$ with three different recruitment functions $f(0,t)$ and initial age structures $f(\alpha,t_1)$ where t_1 is the first sample time. The models were solved numerically by application of the fourth order Runge–Kutta method to the characteristics of the surface $f(\alpha,t)$. Characteristics were calculated every 0.1 'days', and were sampled whenever they crossed a sampling time, which occured every 4 'days'. Each dataset covered about 100 'days' in time and approximately 30 'days' of age. Below we give the functions required to simulate the three datasets.

<u>Set 1</u>:

$$\mu(\alpha,t) = 0.05 + t/500 + 0.6\exp\{-(\alpha-15)^2/50 - (t-40)^2/150\}$$
$$f(0,t) = 1000 + 4000\exp\{-(55-t)^2/150\} + 2000\exp\{-(25-t)^2/100\}$$
$$f(\alpha,t_1) = 1004\exp(-0.05\alpha)$$

Figure 5.5 shows the population and death rate surfaces for this model. The *per capita* death rate increases with time and there is a burst of mortality at age 15 and time 30. This could reflect, for example, a slowly increasing population of invertebrate predators feeding on all age classes and a short bloom of more selective predators, perhaps fish larvae before moving on to bigger prey.

<u>Set 2</u>:

$$\mu(\alpha,t) = 0.03 + 0.3\exp(-y^2/50)/[1 + (5\exp\{-0.02x\}-4)]$$
where $x = \alpha/2 + t - 52.5$ and $y = t - 2\alpha - 45$.

$$f(0,t) = \begin{cases} 500 + 2000\{1-\cos(\pi t/20)\} & t<40 \\ 500 + 4000\{1-\cos(\pi t/20)\} & 40<t<80 \\ 500 & t>80 \end{cases}$$

$$f(\alpha,t_1) = 500\exp(-0.03\alpha)$$

The population and death rate surfaces for this model are shown in figure 5.6. The *per capita* death rate function in this model is supposed to reflect a population of growing predators whose prey size preference increases more slowly than the copepods develop.

<u>Set 3</u>:

$$\mu(\alpha,t) = 0.2\exp(-0.06\alpha) + 0.3\exp\{-(\alpha-15)^2/100 - (t-75)^2/150\}$$
$$f(0,t) = 500 + 3000/[1+0.01(t-25)^2] + 6000/[1+0.007(t-60)^2] + 2\exp(0.05t)$$
$$f(\alpha,t_1) = 1145\exp\{-3.3[1-\exp(-0.06\alpha)]\}$$

The population and death rate surfaces for this last model are shown in figure 5.7. The *per capita* death rate is in this case a decaying function of age with a late mortality burst.

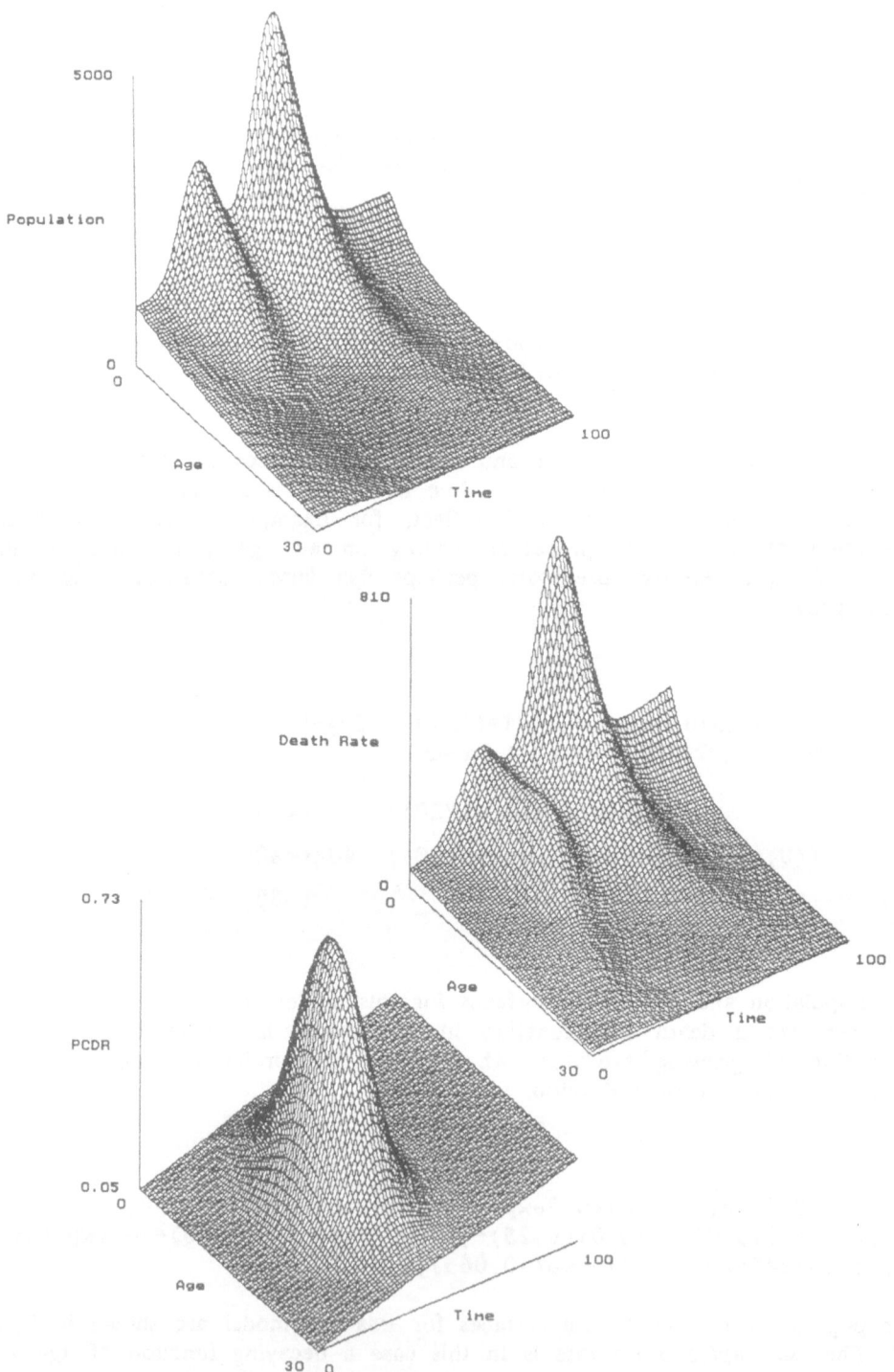

Figure 5.5 The first set of population, death rate and *per capita* death rate surfaces used to test the new method.

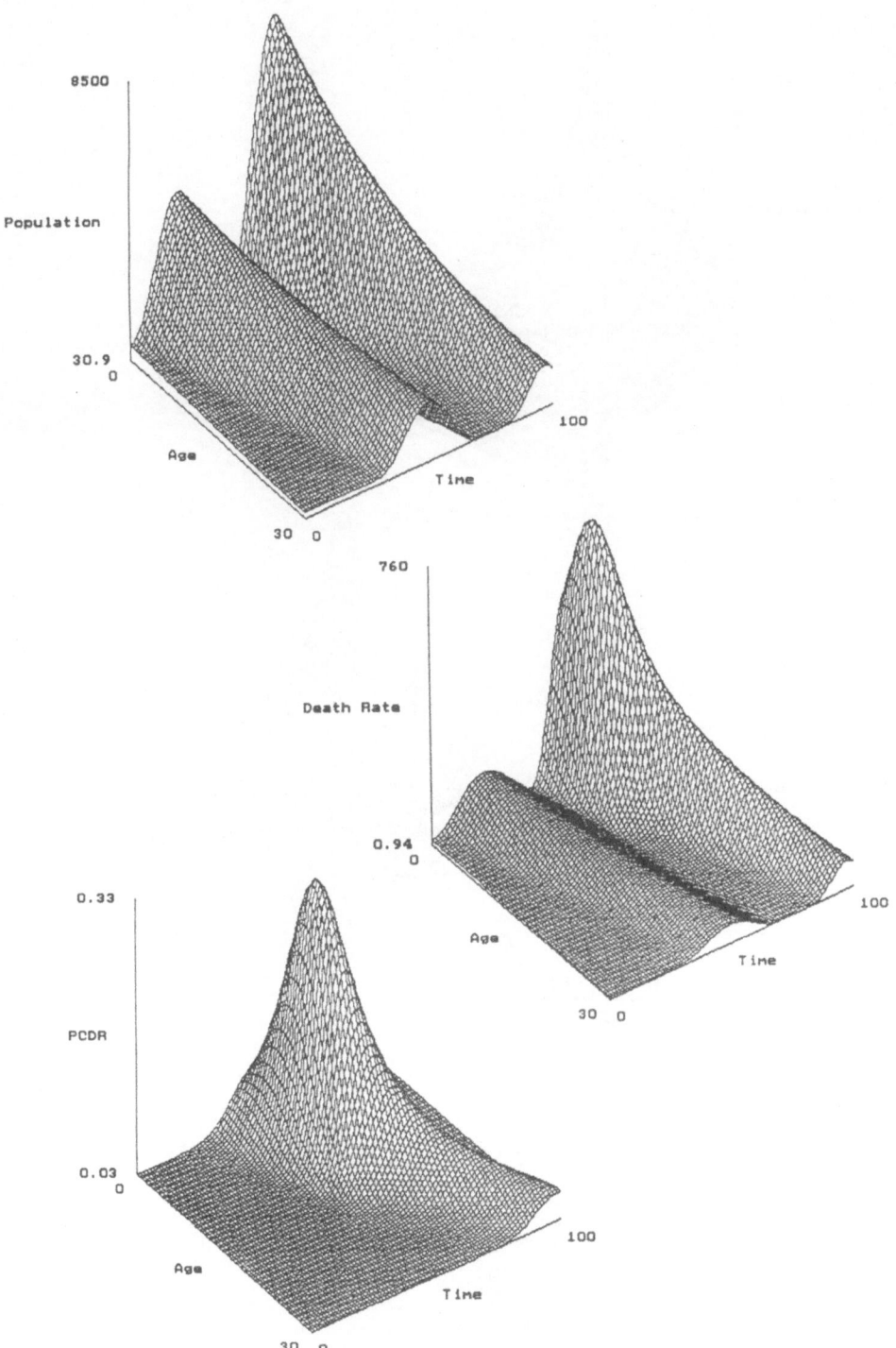

Figure 5.6 The second set of population, death rate and *per capita* death rate surfaces used to test the new method.

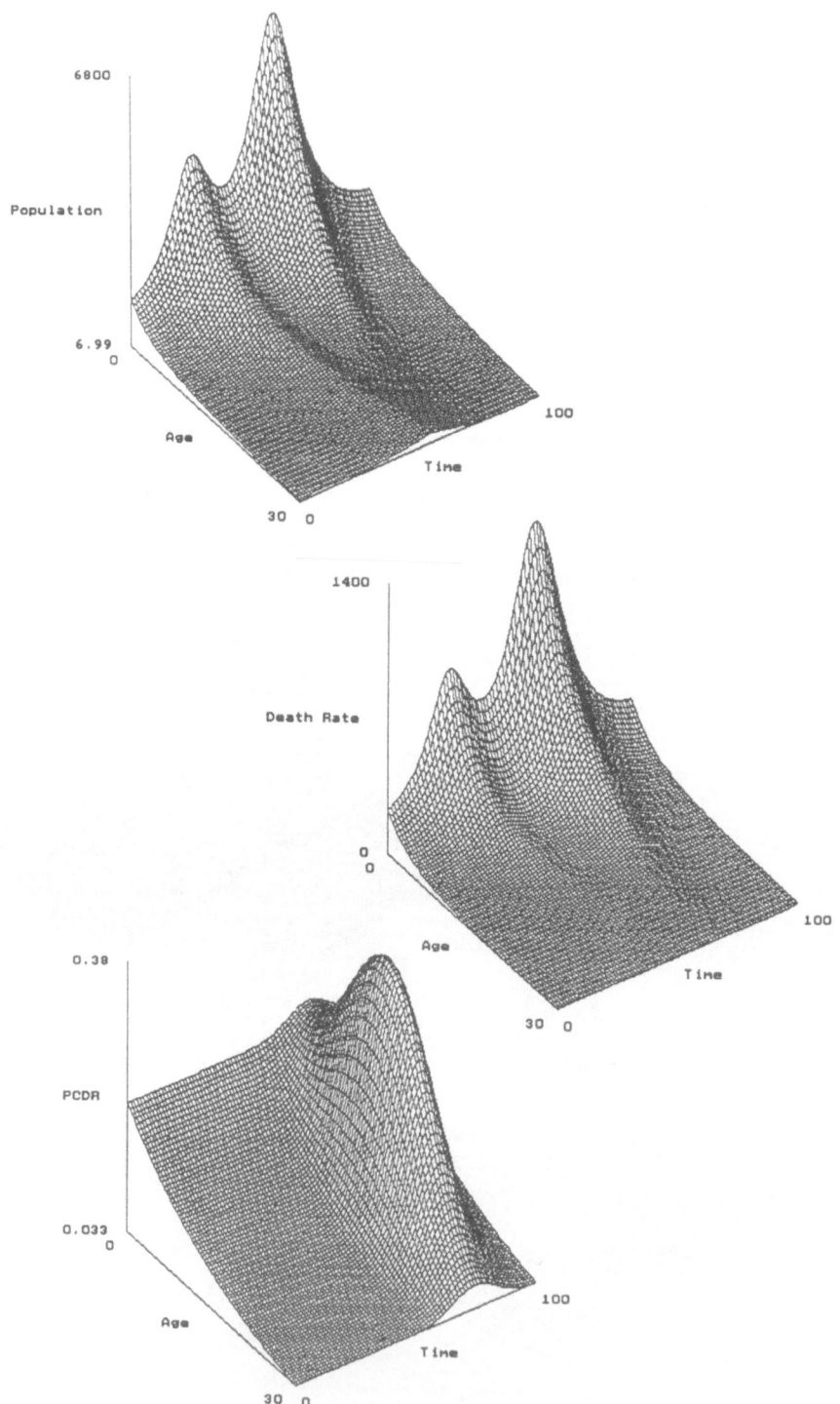

Figure 5.7 The third set of population, death rate and *per capita* death rate surfaces used to test the new method.

All the datasets were perturbed by gaussian noise with stage–specific standard deviation of 10%, 20%, 40% or 60% of the stage population mean. For each of the 12 combinations of model and noise level 10 replicate datasets were produced. The simplified method was applied to all 10 of these and the sophisticated method to two of them.

Results

For each application of a method to a dataset the RMS deviation of the estimated population and death rate surfaces from the 'true' surfaces was calculated as a percentage of the mean of the relevant surface. The RMS deviation of the stage specific death rate (again as a percentage of mean death rate) was also calculated.

t–tests comparing mean RMS deviations for the two methods failed to find significant differences at the 5% level (or indeed the 40% level!) between the means, either for the surfaces or the stage rates for any of the 12 model–noise combinations. However assuming the null hypothesis that on any given dataset either of the two methods had equal probability of doing better than the other method, then the sophisticated method did better than the simplified method at estimating stage specific death rates significantly more times ($p=0.03$). Again there was no significant difference between the number of times that either method did better than the other estimating population or death rate surfaces.

The mean RMS deviations for the population surfaces, death rate surfaces and stage–specific death rates are given in tables 5.10, 5.11 and 5.12 respectively.

Noise level	Model		
	1	2	3
10%	10 (10)	7 (6)	12 (13)
20%	18 (18)	12 (12)	21 (21)
30%	36 (42)	19 (19)	37 (37)
40%	35 (52)	30 (34)	39 (53)

Table 5.10   The means of the RMS deviations of the population surface estimates from the true surfaces, expressed as percentage of mean population.   Figures in brackets are for the sophisticated method, others for the simplified method.

Noise	Model		
level	1	2	3
10%	39 (35)	37 (42)	58 (55)
20%	53 (52)	45 (47)	73 (74)
30%	73 (72)	69 (80)	113 (103)
40%	68 (71)	104 (121)	105 (122)

Table 5.11   The means of the RMS deviations of the death rate surface estimates from the true surfaces, expressed as percentage of mean death rate.   Figures in brackets are for the sophisticated method, others for the simplified method.

Noise	Model		
level	1	2	3
10%	23 (19)	34 (38)	34 (35)
20%	31 (30)	44 (44)	46 (40)
30%	49 (49)	62 (66)	63 (71)
40%	55 (58)	97 (100)	80 (75)

Table 5.12   The means of the RMS deviations of the stage–specific death rate estimates from the true surfaces, expressed as percentage of mean death rate.   Figures in brackets are for the sophisticated method, others for the simplified method.

Figure 5.8 shows some randomly chosen results from the tests. 5.8c is interesting in that it shows how the technique can sometimes miss genuine cohorts altogether, if the data is noisy: in this case standard deviation of the error was 40% of the stage population mean. In other replicates the cohorts *were* identified.

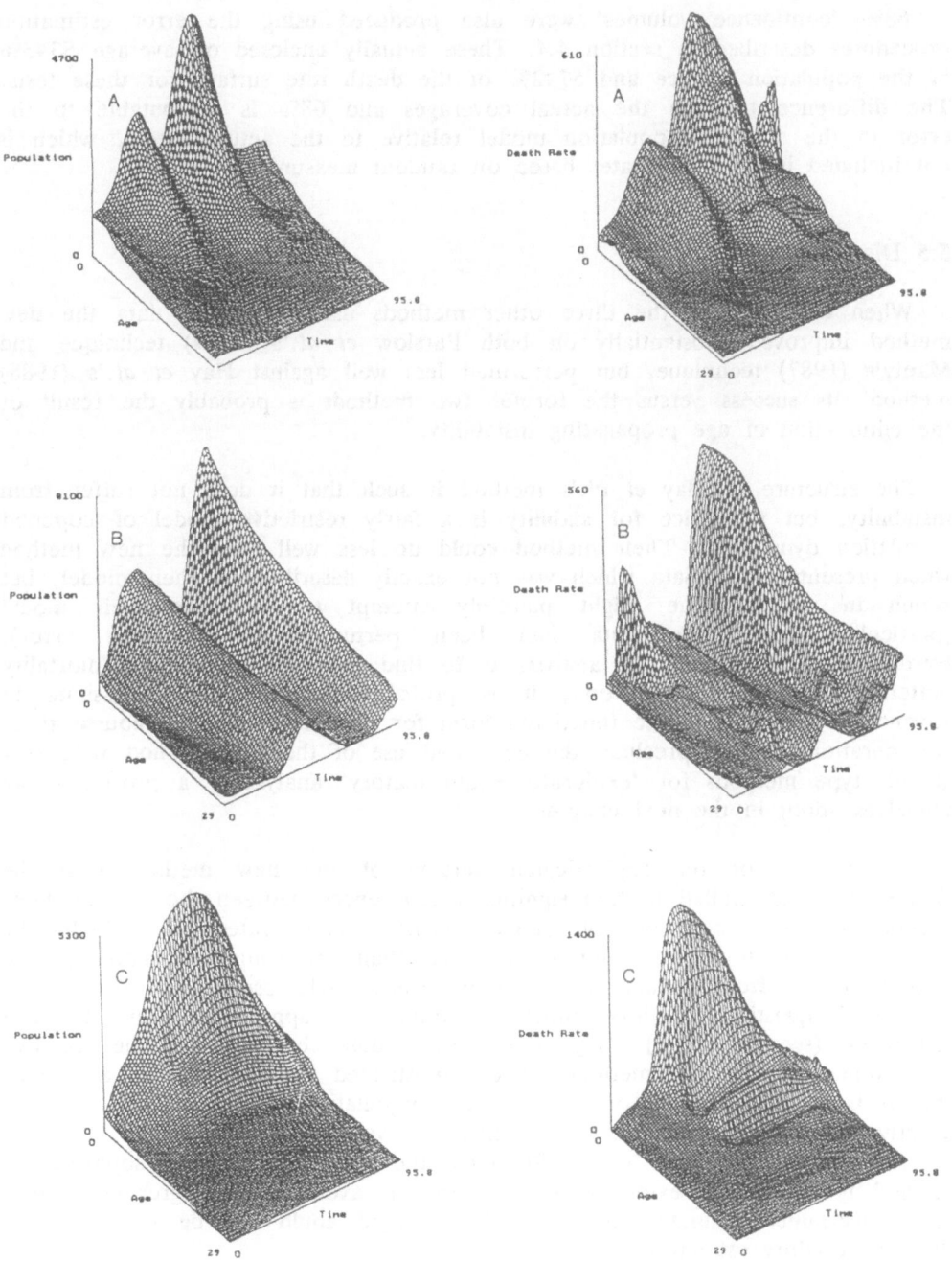

Figure 5.8 Reconstructions of population and death rate surfaces. A) The sophisticated method applied to dataset 1 perturbed by gaussian error with stage-specific standard deviation of 10% of stage mean. B) The simplified method applied to dataset 2 perturbed by 20% noise. C) The sophisticated method applied to dataset 3 perturbed by 40% noise.

68% 'confidence volumes' were also produced using the error estimation procedures described in section 4.4. These actually enclosed on average 53∓3% of the population surface and 57∓2% of the death rate surface for these tests. The difference between the actual coverages and 68% is attributable to the error in the assumed population model relative to the actual model, which is not included in error estimates based on random measurement error.

5.5 Discussion

When compared to the three other methods using simulated data the new method improved substantially on both Parslow *et al.*'s (1979) technique and Manly's (1987) technique, but performed less well against Hay *et al.*'s (1988) method. Its success versus the former two methods is probably the result of the elimination of age–propagating instability.

The structure of Hay *et al.*'s method is such that it does not suffer from instability, but the price for stability is a fairly restrictive model of copepod population dynamics. Their method could do less well than the new method when presented with data which was not exactly described by their model, but which in practice one might plausibly attempt to fit using their model (particularly once the data had been perturbed by sampling error). Furthermore, if the aim of analysis is to find the structure of the mortality patterns affecting a population, it is probably unreasonable to assume in advance a particular simple functional form for the death rate. Of course these considerations do not preclude the combined use of the new method with Hay *et. al.* type methods for 'exploratory–confirmatory' analysis – a procedure we ourselves adopt in the next chapter.

Comparison of the sophisticated version of the new method with the simplified version failed to find significant differences between the two methods except in the estimation of stage specific death rates, in which the sophisticated method did better more often than the simplified method. The tests were far from exhaustive and more work could usefully be done. The unresolved question of how much smoothing is appropriate for the age structures (section 4.6.2) may well have some bearing on the relative effectiveness of the two methods. The sophisticated method has the advantage that it is at least capable of reconstructing population and death rate surfaces exactly, given noise free data; the simplified version of the method can only do this in a few trivial cases. Similarly the formalism of the sophisticated method is capable of extension to find various averaged or aggregated values with uncertainty estimates: the simplified method could not be used to find these uncertainty estimates.

To sum up: the relative merits of the two versions of the new method are as yet largely unknown, but the new method appears to have several advantages over previous methods, in that it can produce reasonable estimates of death rate for a wide range of population dynamics, accompanied by useable uncertainty measures.

CHAPTER 6

LOCH EWE COPEPODS: SOME SPECULATION

Written in collaboration with S.J.Hay
Department of Agriculture and Fisheries for Scotland,
Marine Laboratory,
Victoria Road,
Aberdeen AB9 8DB,
Scotland.

As noted in the Introduction to this monograph, our interest in problems of mortality estimation derived from a number of studies of marine copepods. In particular, we were interested in results from a series of mesocosm experiments performed by members of the DAFS Marine lab, Aberdeen in the 1970s and early 1980s. Interactions between the component populations of complex marine communities were studied by enclosing these communities in large plastic bags suspended in the Loch. In this way systems could be studied which were of far greater complexity than those which might feasibly be reconstituted in an aquarium and the enclosed systems were subjected to a physical environment similar to that found in the rest of the Loch. At the same time the system was closed making accurate sampling of species and environment easy in comparison with the open sea. For a survey of the mesocosm approach see Grice and Reeve (1982).

In this chapter, we consider one series of experiments, performed in 1980 with the objective of identifying factors important to the growth, development and survival of larvae of the Atlantic herring *Clupea harengus harengus*. Some 310 m^3 of loch water was enclosed in each of four cylindrical enclosures, approximately 19.5 metres deep and made of translucent PVC. The enclosures were stocked with extra plankton obtained from the loch and with artificially fertilised herring eggs. The experiments ran for approximately three months, starting in mid-April, but in what follows we will define our "time" variable as Julian day: day 1 is January 1st 1980. Copepod populations were sampled every 4 days and herring larvae at approximately weekly intervals. The bags received no (known) new material from day 107 onwards. Details of the experimental procedures are given in Hay *et al.* (1988) and Gamble *et al.* (1985); our concern in this chapter will be with the dynamics of the three copepod species *Temora longicornis, Acartia clausi* and *Pseudocalanus elongatus* in one particular enclosure (C2: see Hay *et al.* 1988). We shall use these populations as a case study to illustrate the biological judgements and methodological snags associated with applying the new method of mortality estimation to real data.

6.1 Stage durations

The new method of mortality estimation requires prior estimates of stage durations. Since these were not measured *in situ*, there are only two ways of obtaining sets of values: taking values from laboratory experiments and assuming they remain valid in the field, or making use of one of the many indirect methods (e.g. systems identification) which themselves involve major assumptions about the time dependence of mortality rates. In short, a biological leap-in-the-dark or circular logic!

Faced with this choice, we opt for the use of laboratory data, but supplement this with an investigation of the sensitivity of our results to the assumed stage durations. The need for this is established by considering the results of an analysis (using a sytems identification method modified to bias stage duration estimates towards laboratory values – see section 5.2) of the same 1980 Loch Ewe data by Hay *et al*. (1988) who argue that copepod growth rates may be highly flexible. This suggestion is based on large differences in estimated development times for copepodite stages 7 to 11 of *Temora* and *Acartia* between two of the bags, where estimates were considered to be reliable. Nauplii (stages 1 to 6) showed greater consistency in development rate estimates between bags. However we are uncertain about the extent to which these results are attributable to the penalty constraint towards published development rates, which Hay *et al*. adopted in their analysis of both nauplii and copepodites.

Support for the possibility of extrapolating from laboratory data comes from other authors (Corkett and McLaren 1978; Davis 1984) who have argued that in the field food levels are rarely low enough to affect development rate and that temperature is the main factor determining stage durations (see also Thompson 1982 and Klein Breteler *et al*. 1982). One problem with these conclusions is that the arguments supporting them are based on quantity of food, but do not consider food quality. Copepods are able to discriminate and select the food they eat (e.g. De Mott 1988), and food quality has been implicated in the retardation and arrestment of growth and development in field and laboratory copepod populations (Diel and Klein Breteler 1986).

We analysed the data for *Pseudocalanus* in bag C2 using four different sets of published stage durations (see Table 6.1). Fig. 6.1 shows the population surface for each set of stage durations, obtained using the simplified version of the new method. Fig. 6.2 shows the corresponding total death rate surfaces. Only the results obtained using the values from Sazhina (1968, reported in Corkett and McLaren 1978), differ substantially from those produced using the other sets, probably as a result of the large differences in naupliar stage durations between Sazhina and the others.

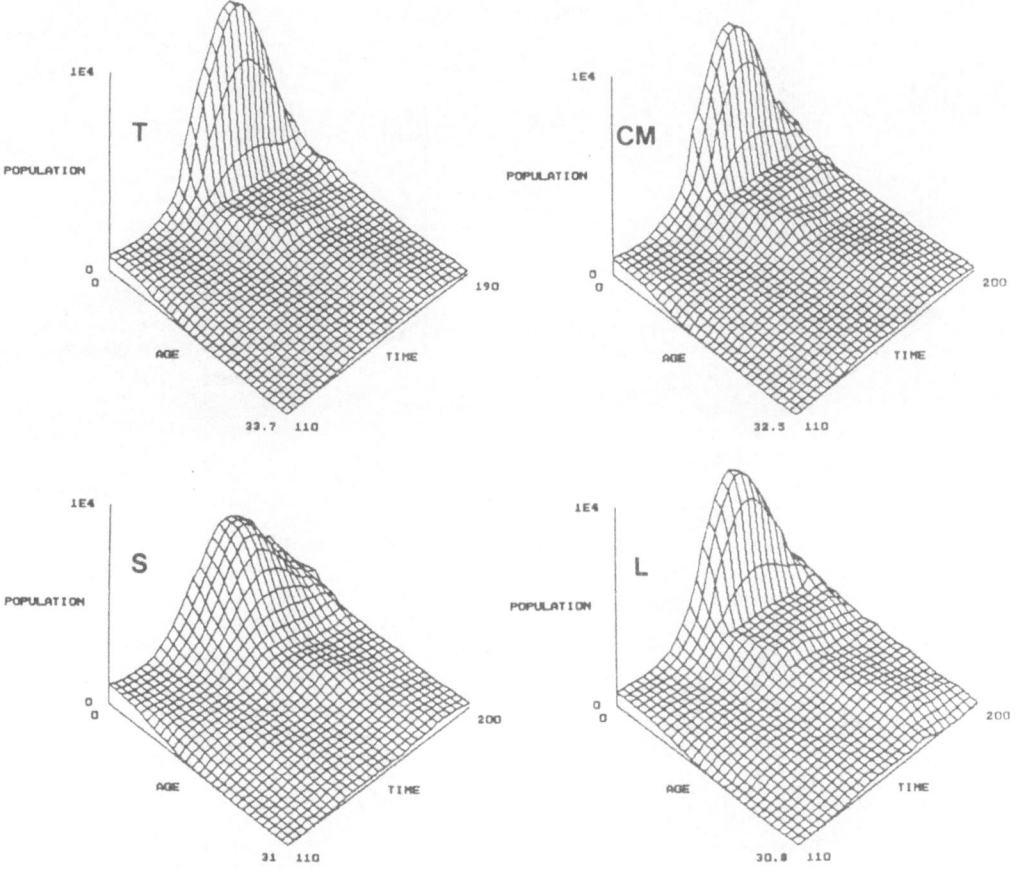

Figure 6.1 Population surfaces for *Pseudocalanus* in a Loch Ewe enclosure (1980: bag C2) calculated using each of the sets of stage durations in Table 6.1. T: Thompson (1982), L: Landry (1983), S: Sazkina (1968), CM: Corkett and McLaren (1978).

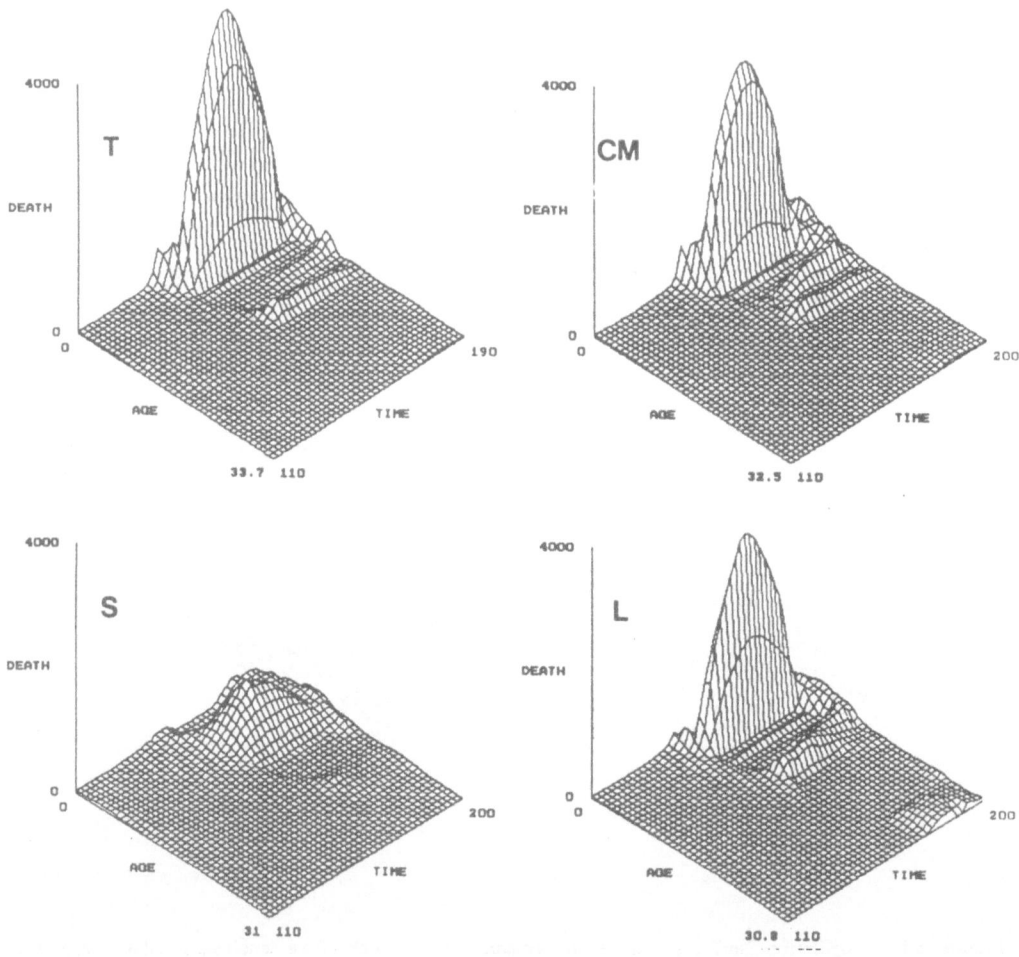

Figure 6.2 The death rate surfaces corresponding to the population surfaces in Fig. 6.1.

Stage	Pseudocalanus				Acartia	Temora
	T	L	S	CM	L	M & K
NI	0.6	0.8	1	0.6	1.0	2.9
NII	1.1	1.3	2	1.3	3.8	2.9
NIII	5.3	4.9	3	5.0	2.2	2.9
NIV	2.8	2.9	3	2.8	1.6	2.9
NV	2.6	2.1	3	2.4	2.4	2.9
NVI	1.2	2.2	2	1.4	3.3	2.9
CI	4.6	3.3	2	4.2	2.9	2.9
CII	3.6	3.5	2	3.4	2.8	2.9
CIII	3.5	2.8	3	3.4	2.5	2.9
CIV	3.7	2.5	5	3.5	4.0	2.9
CV	4.7	4.5	5	4.5	5.5	2.9

Table 6.1   The stage durations (in days) at 10°C used in analysis of the 1980 bag C Loch Ewe *Pseudocalanus* data. The letters heading each column give the source. T: Thompson (1982); L: Landry (1983); CM:Corkett and McLaren (1978); S: Sazhina (1968); K: Klein Breteler *et al.* (1982); M: McLaren (1978).

The results of such a test are open to the critisim that the major differences in stage duration between the sets of durations are in stages N3, N5, N6, C1 and C2. Given that this may be a reflection of either copepod biology or small sample size or both, we decided to apply a more general test of the method's sensitivity to stage duration estimates.

We assumed that the variability in stage duration as a fraction of mean stage duration could be described by a single normal distribution applicable to all stages. The standard deviation of the distribution was estimated from the data in table 6.1 and found to be about 20% of the mean stage duration. Routines given in Press *et al.* (1986) were used to generate 100 sets of normally distributed simulated stage durations. The simplified version of the method of chapter 4 was applied to the *Pseudocalanus* data (bag C2, 1980) using each of the duration sets.

Direct calculation of 95% confidence intervals is complicated by the fact that stage duration changes alter the apparent position of death rate features in the age–time plane. This is unimportant biologically, since we cannot tell whether a death happened (for example) at the end of stage i or the

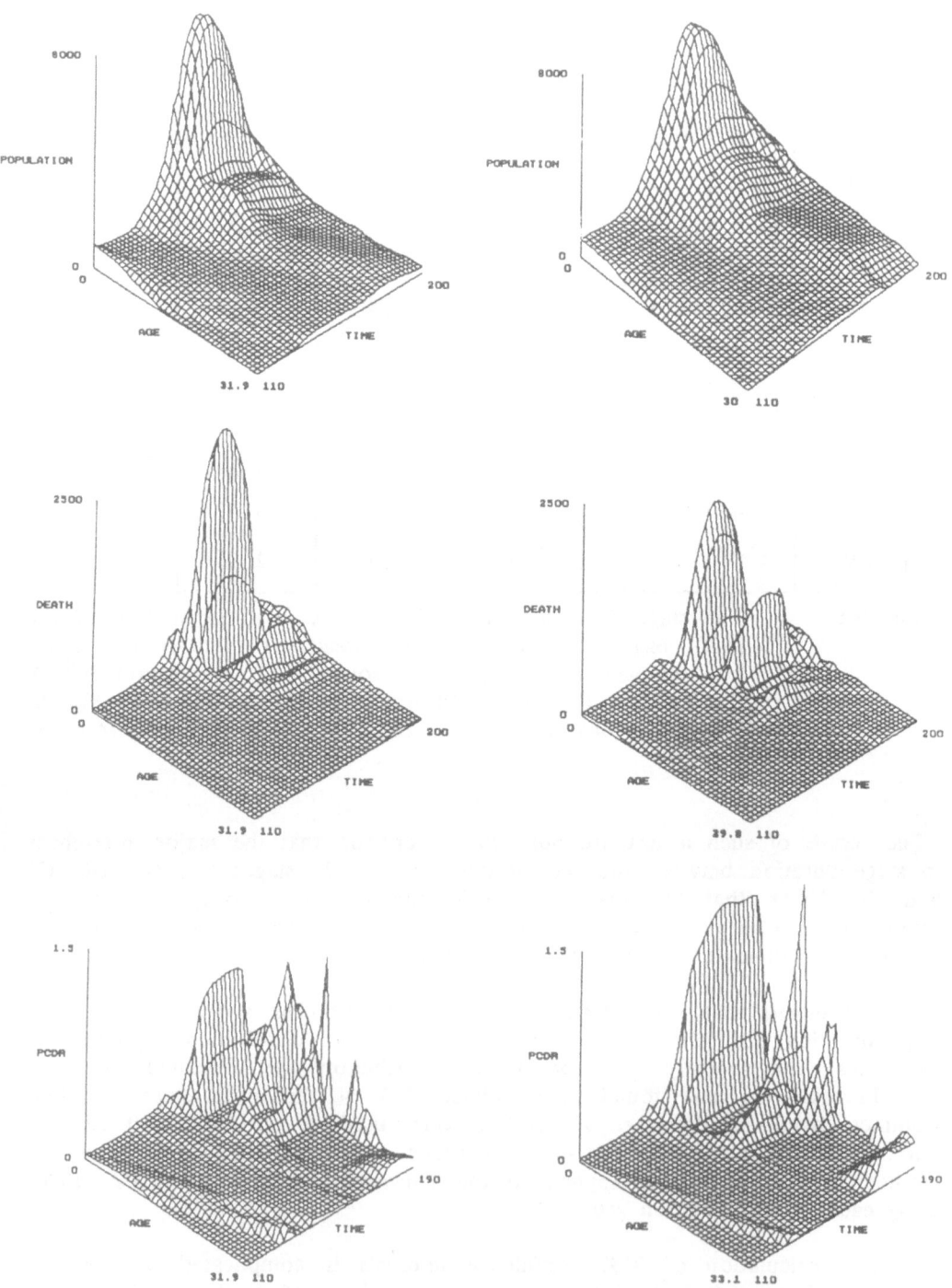

Figure 6.3 "Mean surfaces" (left hand column) and "5 per cent results" (right hand column) for the population of Fig. 6.1. Also shown are total and *per capita* death rate surfaces.

beginning of stage i+1, but such shifts will totally dominate simplistic attempts to derive confidence intervals by Monte–Carlo simulation.

For these reasons we did not calculate confidence intervals directly. Instead, each of the 300 death rate, population and *per capita* death rate surfaces were compared with the corresponding surfaces calculated from the arithmetic mean durations implied by table 6.1 (henceforth, 'mean surfaces'). The surfaces were ranked according to mean square deviation from the appropriate mean surface. Figure 6.3 compares each of the surfaces having the 5th largest deviation from the mean surfaces, to give some idea of the 95% point on the error distribution resulting from stage duration variability. Figure 6.4 shows the population and death rate surfaces from figure 6.3, rescaled to compare detail away from the peaks in population and death rate.

The two figures demonstrate that the qualitative features of the death rate and population surfaces are robust to variability in assumed values for stage durations. Furthermore the size of the major features of the surfaces are not greatly changed by changes in stage duration of an order consistent with table 6.1. The most obvious conclusion for *Pseudocalanus* in the bag is unarguable: most deaths occur towards the end of the summer in the nauplii stages.

Bearing in mind the *Pseudocalanus* results, we turn to the other two species. Landry (1983) published stage durations for *Acartia tonsa* for all stages at 15˚C and for the copepodite and final nauplii stages of *Acartia clausi* at the same temperature. He also published the total duration of the nauplii stages up to 5 for *A. clausi* and durations for *Pseudocalanus* at 15˚C. To obtain the missing nauplii durations for *A. clausi* we assumed that the ratio of total development time to N6 to stage duration was the same for each of the first 5 stages of *A. clausi* and *A. tonsa*. The stage durations at 15˚C were then converted to estimates of durations at 10˚C by scaling all the durations in the same ratio as the total development times of *Pseudocalanus* at 15˚C and 10˚C (following the hypotheses on temperature dependance given in Corkett and McLaren, 1978). The final estimates are given in table 6.1.

For *Temora* the results of Klein Breteler *et al*. (1982) give strong evidence of isochronal development, assuming that food supply is sufficient. Using McLaren's (1978) estimate of total development time we obtain an estimated stage duration for every stage of 2.9 days.

6.2 Modelling sampling error

The distribution of errors affecting any population sample is unknown, so it is prudent at this stage to see what affect different models of the error distribution might have on our results. Three models were considered: In the first it was assumed that each stage population time–series was subject to Gaussian noise with stage dependent variance. This is the case assumed in chapter 4 and is equivalent to setting all the weights w_{ij} in equation (4.2.1) to one. This is also the model assumed by Hay *et al*. (1988) in their analysis of

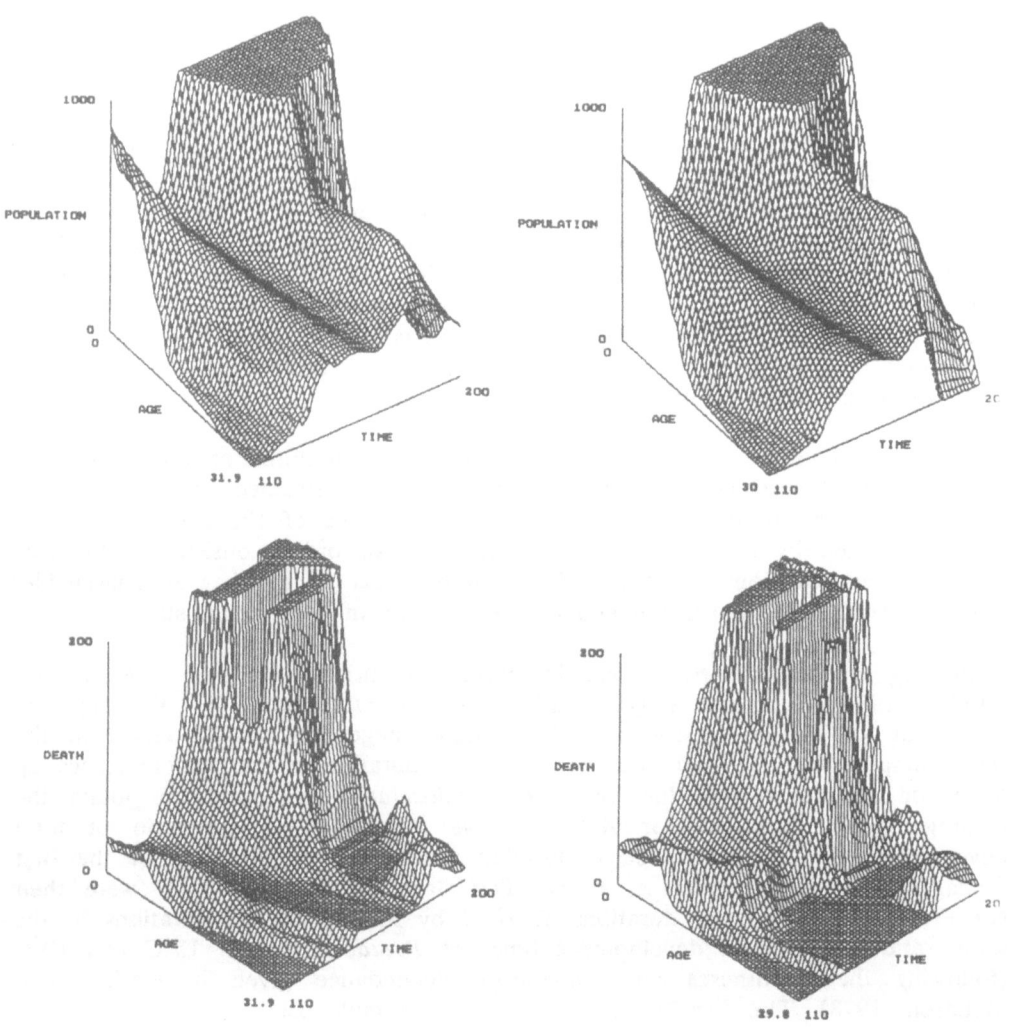

Figure 6.4   Fig. 6.3 but with population axis rescaled to emphasise features away from the population peaks.

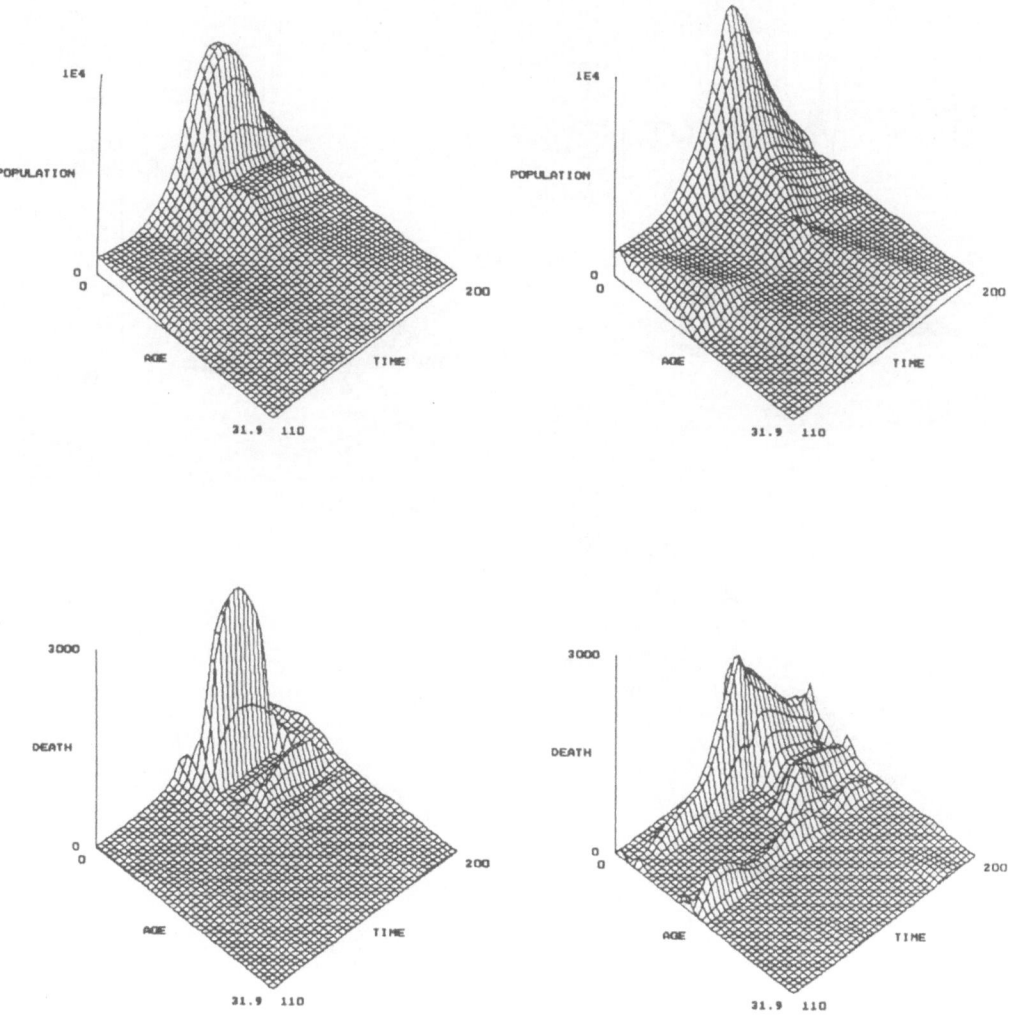

Fig. 6.5 Population and death rate surfaces for the *Pseudocalanus* population of Fig. 6.1 reconstructed using the "normal" error model (left hand column) and the "log-normal" error model (right hand column).

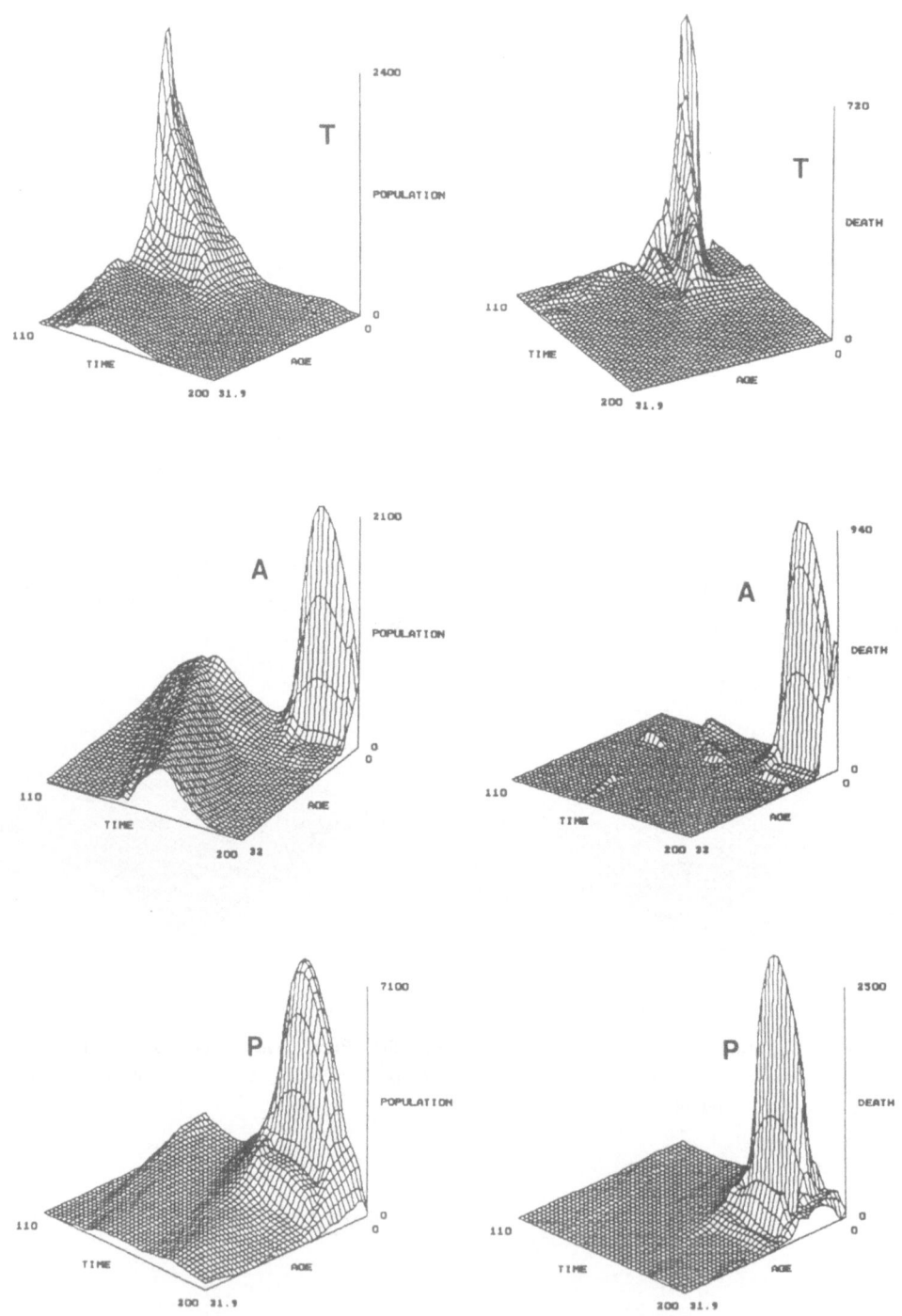

Fig. 6.6 Population and death rate surfaces for *Temora* (T), *Acartia* (A), and *Pseudocalanus* (P) in the Loch Ewe enclosure (1980: bag C2).

these data. In the second model it was assumed that variance of the error affecting a datapoint was proportional to the underlying population value. Smoothing splines were fitted to the stage population time series to obtain initial population estimates and these estimates were then used to obtain weights for each datapoint before refitting splines to the time–series. The final error model can only be used with the simple version of the method: the time series were log–transformed before splines were fitted to them. The resulting smoothed time series estimates were then back–transformed (see for example Meyer, 1975 p. 285) and the associated variance estimated by a first order linear approximation.

When applied to the *Pseudocalanus* data the first two models produce almost identical results. The third model produces slightly different population estimates and slightly lower death rate estimates spread over more stages. Figure 6.5 compares the population and death rate surfaces for *Pseudocalanus* using the first (Normal) model and the third (Log–Normal) model. Again the result that the deaths are concentrated in the naupliar stages late in the experiment is robust.

6.3 Death rate patterns for the different copepod species in bag C2: comparison and speculation.

Having looked at the problems we now look at the data. Figure 6.6 shows the death rate and population surfaces which result from applying the simple version of the new method to the data for *Temora*, *Pseudocalanus* and *Acartia* assuming a normal error model. The actual time–series and the final stage population estimates from them are shown in figures 6.7 to 6.9.

There appears to be a succession in which *Temora* starts out dominant but suffers heavy mortality in the nauplii stages, *Acartia* has a small population with an initial peak after the decline of *Temora* and some evidence of increased death rate at about this time. There is an early cohort of *Pseudocalanus* which appears to suffer little mortality, but the major peak in *Pseudocalanus* numbers appears towards the end of the experiment, and heavy mortality is suffered in the nauplii stages, with only the early part of this second 'cohort' surviving to the copepodite stages. Finally when *Pseudocalanus* numbers are in decline there is a final burst of *Acartia* nauplii which suffer very heavy mortality and do not get beyond the early stages. The same general picture emerges from analysis using the Log–Normal error model.

All three species suffer heavy naupliar mortality in at least one cohort, but the mechanisms may be different in *Temora* as opposed to the other two species. Mortality in the early *Temora* cohort occurs in the later larger nauplii whereas the second cohorts of both *Pseudocalanus* and *Acartia*, which peak towards the end of the experiment, suffer heavy mortality in the very young stages. First feeding in copepods probably occurs in naupliar stages 2 or 3 (2 for *Acartia* and 3 for *Pseudocalanus*) and the rapid decline of the youthful *Pseudocalanus* cohort and even more precipitous demise of the slightly later

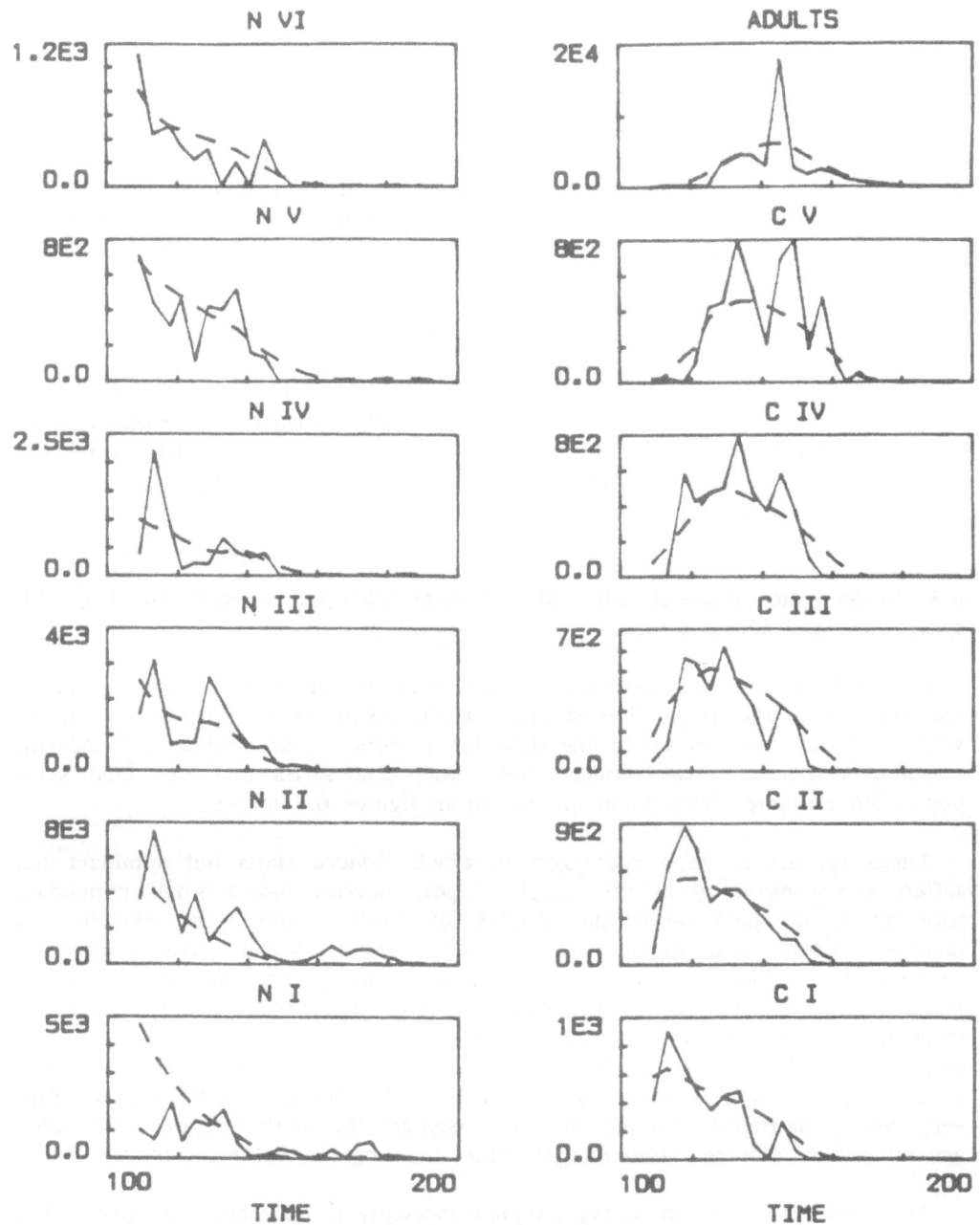

Fig. 6.7 Measured time series (continuous line) compared with inferred stage populations (broken line) from Fig. 6.6 for *Temora*.

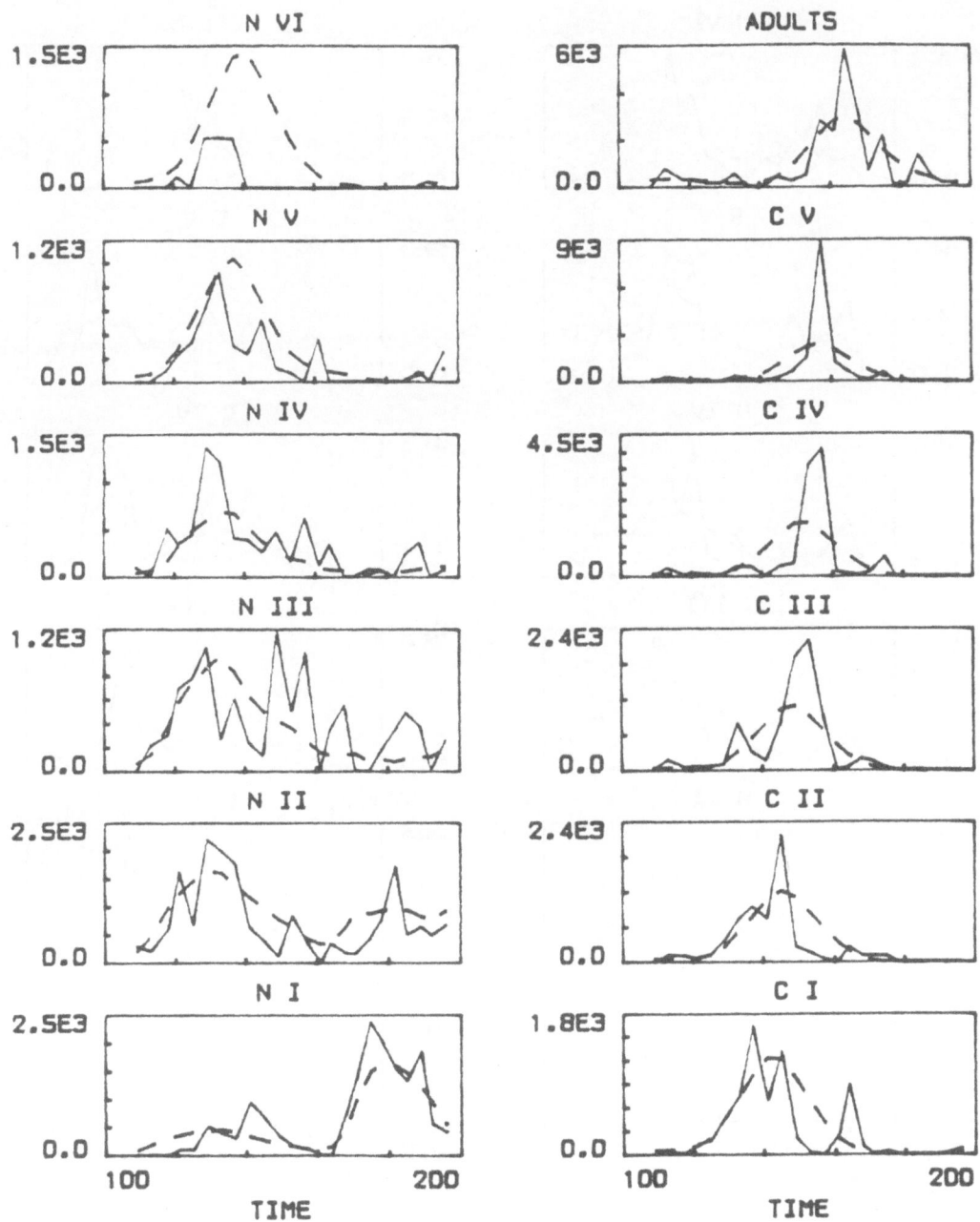

Fig. 6.8 Measured time series (continuous line) compared with inferred stage populations (broken line) from Fig. 6.6 for *Acartia*.

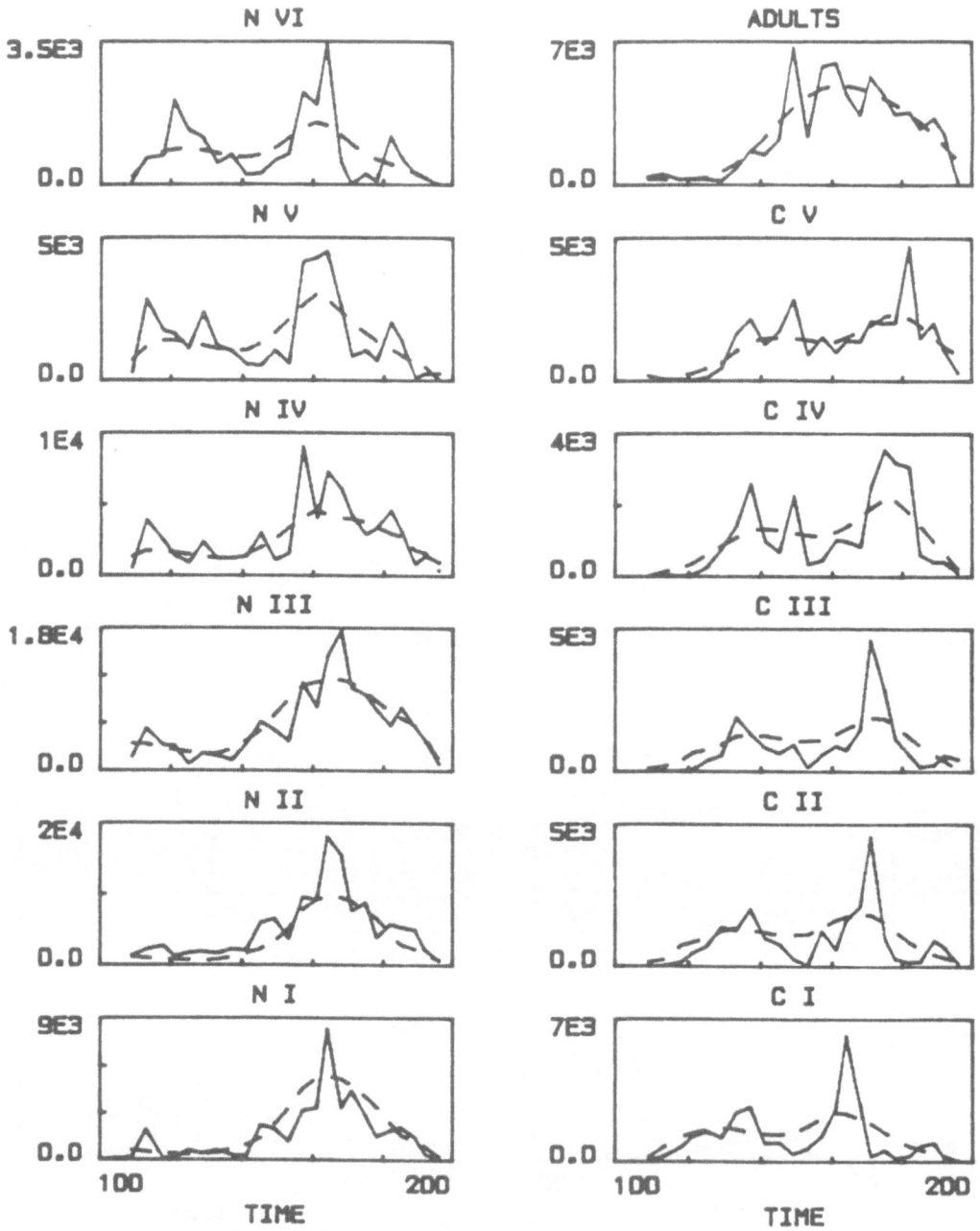

Fig. 6.9 Measured time series (continuous line) compared with inferred stage populations (broken line) from Fig. 6.6 for *Pseudocalanus*.

cohort of still younger *Acartia* is suggestive of a worsening food shortage late in the experiment. Nauplii of *Temora* don't start dying until they have been feeding for some time. This suggests that they fall prey to the principal predators in the enclosure: herring larvae.

Food shortage alone is not the only candidate explanation for the decline in the later cohorts of *Acartia* and *Pseudocalanus*. It is known that the late copepodite stages of both *Acartia* and *Temora* will feed on young nauplii stages. Compare in Fig. 6.4 the variation in numbers of late copepodite *Temora* through time with the variation in time of the death rate of early nauplii of *Acartia* in the middle of the experiment. Now compare the timing of the peak in late stage *Acartia* with the peak in *Pseudocalanus* death rate. Such coincidences suggest that interspecific predation among copepods may have occurred.

Gamble *et. al.* (1985) observed that copepod nauplii formed a substantial part of the diet of the Herring larvae in the bags and obviously a careful linking of our results to stomach analysis is desirable. Unfortunately we don't have equivalent data for copepod diet! Let us speculate anyway: The first major peak in *Temora* suffers heavily from predation by Herring larvae whilst *Pseudocalanus* and *Acartia* escape relatively unscathed. Because the cohort of *Temora* started early those individuals which did survive were able to exploit part of the emerging cohort of *Acartia*. This same cohort of *Acartia* is later poised to exploit the second cohort of *Pseudocalanus* in lieu of declining phytoplankton density and hence to produce a second cohort, but this is too late for the food supply. Such speculation is of course an attempt to solve another inverse problem: what is the community model which explains the death rate and population patterns that we have obtained? Splines, regularization or smoothing aren't going to help us here! Careful analysis of the other Loch Ewe replicates, stomach contents data and phytoplankton data are what is needed. Another useful extension would be biomass plots which could easily be produced from our analysis and might improve our understanding of energy flow within the bags.

Clearly the new method is capable of facilitating further exploration of the Loch Ewe data by providing insights and speculations that may suggest new lines of investigation and future experiments.

CHAPTER 7

DISCUSSION

7.1 What we've done

In the preceding chapters we identified instabilities which largely explain the problems that bedevil mortality estimation methods for stage structured populations. We then showed how these problems can be overcome using the biologically sensible assumption that mortality rate varies slowly over the development time of a cohort. The proposed method fits stage structured population data with a solution of the McKendrick–Von Foerster equation based on spline functions. On the basis of computer simulation studies this method appears to offer substantially better results than 'systems identification' or the method of Manly (1987) and preliminary results with real data are encouraging.

The implementation of our method is ugly (sections 4.6 to 4.7), but it works and hopefully demonstrates that sensible choice of modelling assumptions can be used to produce meaningful death rate estimates. Indeed if we had not been able to assume smooth variation of death rate then there would have been no way of deciding which members of an unpleasantly large family of mortality and birth functions were responsible for our datasets.

This feature of requiring some form of smoothness in the solution is common to inverse problems from physics (e.g. Tikhonev, 1963) to neural networks (Poggio and Girosi, 1989) and includes such esoterica as drawing a straight line through a set of datapoints. Caswell and Twombly (1989) have recognised this need for smoothness and solved a parameter estimation problem based on matrix models using Tikhonev regularization (Tikhonev, 1963), which is related to spline smoothing (Ragozin, 1983). Those requiring heavy detail on inverse problems can consult Talenti (1986) or Morozov (1984). The readers digest version comes in the slightly tangential but excellent Poggio and Girosi (1989). They point out that neural networks are effectively solving hypersurface reconstruction problems using assumptions of smoothness, in a manner closely related to generalised spline theory. Ecological parameter estimation is of course a special case of a hypersurface reconstruction problem and the neat thing about neural networks is that they provide a *simple* algorithm for solving such problems. Ecological inverse problems could be solved by training a network to recover parameters from simulated datasets and then applying the trained network to real data. There are three obvious problems with this approach. One still needs (if indirectly) to control the trade off between generality of the estimator across datasets with fidelity in reconstruction of individual test datasets, the problem which we solve using cross–validation. The second problem is that the simulated datasets on which a network could be trained need in some sense to span the space of possible real datasets (this is related to the first problem!). The third problem is that given current computer hardware, neural networks are expensive to train – this is a problem which may vanish with time.

A final general point relates to the frequent assertion that sampling intervals should always be less than stage durations for successful parameter estimation. This isn't true. It is true that some methods can become ill–conditioned if the condition is not met, but the problem can be overcome, in the manner of most inverse problems, if smoothness can be assumed so that interpolation between datapoints is possible. Sometimes sampling will be too infrequent to allow successful population trajectory reconstruction by interpolation, but what matters then is the timescale of recruitment and mortality rate variation relative to sampling interval. The stage duration is not really relevent. It may be that for some species, stage durations give a useful rule of thumb for choosing a sampling interval, but for a few copepod species at least, even this doesn't seem to be true.

7.2 Experimental Suggestions

Anyone who has ever tried to sample, sort and count plankton will be aware of just what an achievement figure 6.8 represents, however there are at least two means by which the information content of such datasets could be substantially increased.

Firstly replicate sampling of at least a few datapoints would reduce the reliance on cross validation and hence enable shorter experimental runs to be analysed. More frequent sampling is an obvious alternative. The data used in this study was collected using single vertical net hauls, which were then sub–sampled and counted. The combination of small scale patchiness in the enclosure and ordinary sampling error associated with net–hauling and the sub–sampling creates a complicated error proccess. What makes it really nasty is the human error inevitable when sorting and counting organisms whose early stages are very small and quite similar in appearance between species. This latter component of the noise could be investigated without the need for additional net hauls (and hence additional mortality) by replicate sub–sampling of single samples. The priority for future sampling must be non–destructive automated methods of identification and counting. Neural network plankton counters perhaps?

The second improvement concerns stage durations. As Hay *et al.* (1988) recognised, any estimation method will do better if stage durations can be taken as known parameters. Our method requires that stage durations are known and we have used laboratory measurements. Extrapolation from the lab to the loch is controversial: a glass tube of filtered sea water on a constant temperature block under artificial illumination isn't a terribly exact simulation of spring sea conditions on the Scottish coast. The ideal procedure would be measurement in small enclosures *in situ* as reported in Braner and Hairston (1989). This should enable durations to be assessed in environmental conditions as near identical to those pertaining in the experiment as possible.

7.3 Production Estimation

A sensible definition of production is the amount of new biomass produced during a specified time interval (the literature also has plenty of silly definitions). Being able to produce production estimates for different life history stages is important for studies of predation efficiency, optimal foraging, and forces controlling succession. Given a population surface it is easy to produce a surface representing production per unit time by multiplying the population at each point on the surface by the individual rate of increase of mass at that point. This idealised procedure relies on a knowledge of the individual growth rate function and the assumption that it is the same for all individuals.

An alternative method of deriving production estimates involves summing the weights of individuals *dying* over all age classes and a particular time interval, and adding the rate of (total) biomass increase. This estimate derives production from the population and biomass surfaces and avoids the need for knowledge of growth rates. Our method for mortality estimation opens the possibilites of production estimates which do not require the restrictive assumptions about underlying population dynamics which are a feature of many existing production estimation methods (surveyed in Romanovsky and Polishchuk, 1982). In practice, however, it may not be worth the considerable effort involved in implementing our method if it is *only* production estimates that are required from a dataset. Production estimates appear to be fairly robust to the errors in mortality rate estimates which our method combats, so there is no reason to expect the new method to do very much better than much simpler production formulae, except in situations with rapid variations in age structure. A set of tests, similar to those presented in chapter 5, would help identify precisely those situations in which sophisticated estimation methods are required.

7.4 Methodological Improvements

Two general points before moving on to specifics. Firstly, methods to date have not tackled the question of mortality rate variations within stages at a given time. The significance of such variations should be investigated, since mortality rates are probably inflated close to stage transitions. Secondly, distributed stage durations have received a similar lack of attention in the parameter estimation literature, although recent work by Braner and Hairston (1989) on stage duration estimation has gone some way towards rectifying this.

Systems identification again

If assumptions of smoothness are the key to avoiding instability then systems identification might benefit from the inclusion of a death rate smoothness constraint in the objective function used for model fitting. The problem of how much weight to give the smoothing part of the objective function might be soluble by some varient of cross-validation. The major concern is that the smoothing term should not be given so much weight that it causes large changes in the fitted population trajectories, since its role is to pick the most feasible parameter set from the set of all parameter sets yielding similar population trajectories. Decisions on what constitutes a 'similar' parameter set could be based on the confidence intervals associated with fitted population trajectories.

Better surface estimation methods

The extension of our current mortality estimation method to the case of time varying stage durations is conceptually trivial, if computationally offputting. For the loch Ewe data reported in chapter 6 such an extension is desirable given a 4°C temperature variation over the period of the experiment, although the qualitative results are likely to remain much the same.

Much harder will be the tidying of the method from its current three pass implementation to a one pass surface fitting method enabling all cross-validation to be done simultaneously and thereby making maximum use of between and within stage population correlations. This will involve some very costly matrix manipulations, but should also facilitate the inclusion of monotonicity constraints on the characteristics as a problem in quadratic programming: the systems solved to fit smoothing splines are in quadratic form, and Fritsch and Carlson (1980) give sufficient conditions for monotonicity of piecewise cubics which can be expressed as linear constraints on the knots of a spline function.

Estimation of stage durations from existing data is the biggest outstanding problem not covered by surface reconstruction. As we have hinted above, one way forward might be through hypersurface reconstruction with neural networks.

7.5 The Take Home Message

Mortality rate estimation is an underconstrained 'inverse problem' and is therefore prone to instability, but biologically sensible modelling assumptions of smoothness can be used to overcome these difficulties. We have demonstrated this approach using a specific problem and hope that similar techniques will be useful in dealing with other ecological inverse problems.

Appendix : An algorithm for positive distribution estimation.

This appendix outlines the algorithm used to construct figure 3.2. Obviously it is one of a considerable family of possible schemes for producing positive distribution estimates. We refer to the end conditions s=0, s'=0 as type I end conditions and use the notation $\eta(i,j)$ for a dataset running from x_i to x_{j+1}.

1 : Set both end conditions to natural.

2 : Fit $s(x)$ to $\eta(i,j)$ using the method of section 2 and find x_{min} occuring in the kth interval being the position of the minimum of $s(x)$ between x_i and x_{j+1}. If this minimum is less than zero then do (3:) otherwise return $s(x)$ as the fit to $\eta(i,j)$.

3 : If $k = i$ or j then do (4:) otherwise do (5:).

4 : If the end condition for the kth interval is natural then set it to type I and reapply (2:) to obtain $s(x)$ otherwise contract the kth interval to x_{min} without changing $\eta(i,j)_k$ and reapply (2:). The distribution estimate $s(x)$ will be constructed by attaching a short zero section from the end of the estimate returned by (2:) to the end of the original dataset. Return $s(x)$.

5 : Split $\eta(i,j)$ into two sections $\eta(i,k)$ and $\eta(k,j)$. The element $\eta(i,j)_k$ is divided between the two sections in the ratio

$$[\int_{x_k}^{a_{min}} s(x)dx - s(x_{min})(x_{min}-x_k)] : [\int_{x_{min}}^{a_{k+1}} s(x)dx - s(x_{min})(x_{k+1}-x_{min})]$$

Type I end conditions are applied to element k of each section and (2:) is then applied to each. The resulting distribution estimates are combined to yield $s(x)$ which is returned as the distribution estimate.

97

REFERENCES

Asknes D.l. and T.J. Hoisoeter (1987). Obtaining life table data from stage frequency statistics. *Limnol. Oceanogr.* 32: 514-517.

Asknes D.I. and T. Magnesen (1988). A population dynamics approach to the estimation of production of four calanoid copepods in Lindaspollene, western Norway. *Marine Ecology Progress Series* 45: 57-68.

Boneva L.I., D. Kendall and I. Stefanov (1971). Spline transformations: three new diagnostic aids for the statistical data-analyst. *J. Roy. Stat. Soc* 33: 1-77.

de Boor (1978). A Practical Guide to Splines. Springer Verlag, New York.

Braner, M. and Hairston, N.G. (1989). From cohort data to life history parameters via stochastic modelling. *in* Estimation and Analysis of Insect Populations (L. McDonald, B. Manly, J. Lockwood and J. Logan, Eds.), Springer-Verlag Lecture Notes in Statistics, 55: 81-92.

Burden, R.L. and J.D. Faires (1985). Numerical Analysis, 3rd. Edition. Prindle, Weber and Schmidt, Boston, USA.

Burns C.W. (1980). Instar development rates and production of three generations of Boeckella dilatata (Copepoda: Calanoida) in a warm monomictic lake. *Int. Theor. Angew. Limnol. Verh.* 21: 1578-1583.

Caswell, H. and S. Twombly (1989). Estimation of stage-specific demographic parameters for zooplankton populations: methods based on stage-classified matrix projection models. *in* Estimation and Analysis of Insect Populations (L. McDonald, B. Manly, J. Lockwood and J. Logan, Eds.), Springer-Verlag Lecture Notes in Statistics, 55: 94-107.

Comita G.W. (1972). The seasonal zooplankton cycles, production and transformations of energy in Severson Lake, Minesota. *Arch. Hydrobiol.* 70:14-66.

Corkett C.J. and I.A. McLaren (1978). The biology of Pseudocalanus. *Adv. Marine Biol.* 15:2-231.

Craven P. and G. Wahba (1979). Smoothing noisy data with spline functions: estimating the correct degree of smoothing by the method of generalized cross validation. *Numer. Math.* 31: 377-403.

Davis, C.S. (1984). Food concentrations on Georges Bank: non-limiting effect on development and survival of laboratory reared *Pseudocalanus sp.* and *Paracalanus parvus* (Copepoda: Calanoida). *Marine Biol.* 82: 41-46

Dyn N. and G. Wahba (1982). On the estimation of functions of several variables from aggregated data. *SIAM J. Math. Anal.* 13: 134-152.

Edmondson W.T. (1960). Reproductive rates of rotifers in natural populations. *Memorie Ist. Ital. Idrobiol.* 12: 21-77.

Edmondson W.T. (1968). A graphical method for evaluating the use of the egg ratio technique for measuring birth and death rates. *Oecologia* 1: 1-37.

Elden L. (1984). An algorithm for the regulation of ill-conditioned banded least squares problems. *SIAM J. Scient. Stat. Comput.* 5: 237-254.

Evans G.T. (1985). Poster presentation at International Council for the Exploration of the Sea. CM1985/Minisymposium/No8.

Fransz H.G. (1980). Computation of secondary production of Calanus finmarchius using a multiple regression model. *Proc. Final ICES/JONSIS Workshop on JONSDAP '76*. ICES CM 1980/C 3:99-109.

Fransz H.G., J.C. Miquel and S.R. Gonzalez (1984). Mezozooplankton composition, biomass and vertical distribution, and copepod production in the stratified central North sea. *Netherlands Journal of Sea Research* 18: 82-96

Fritsch, F.N. and Butland, J. (1984). A method for constructing local monotone piecewise cubic interpolants. *SIAM. J. Scient. Stat. Comput.* 5: 300-304.

Fritsch, F.N. and R.E. Carlson (1980). Monotone piecewise cubic interpolation. *SIAM J. Numer. Anal.* 17: 238-246.

Gamble J.C., P. MacLachlan and D.D. Seaton (1985). Comparative growth and development of autumn and spring spawned Atlantic herring larvae reared in large enclosed ecosystems. *Marine Ecology Progress Series* 26:19-33.

Grice G.D., and Reeve M.R. (1982). Marine Mesocosms. Springer Verlag New York

Gurney W.S.C., R.M. Nisbet and J.H. Lawton J. (1983). The systematic formulation of tractable single species population models incorporating age structure. *J. Anim. Ecol.* 52: 479-495.

Gurney W.S.C., R.M. Nisbet and S.P. Blythe (1986). The systematic formulation of models of stage structured populations. *in* The Dynamics of Physiologically Structured Populations (J.A.J. Metz and O. Diekmann Eds). Springer Verlag Lecture notes in Biomathematics 68: 475-494.

Harris R.P., M.R. Reeve, G.D. Grice, G.T. Evans, V.R. Gibson, J.R. Beers and B.K. Sullivan (1982). Trophic interactions and production processes in natural zooplankton communities in enclosed water columns. *in* Grice and Reeve (1982).

Hairston N.G., W.E. Walton and K.T. Li (1983). The causes and consequences of sex-specific mortality in a freshwater copepod. *Limnol. Oceanogr.* 28: 614-624.

Hairston N.G. and S. Twombly (1985). Obtaining life table data from cohort analyses: A critique of current methods. *Limnol. Oceanogr.* 30: 886-893.

Hay S.J., G.T. Evans and J.C. Gamble (1988). Birth, growth and death rates for enclosed populations of calanoid copepods. *J. Plank. Res.* 10: 431-454.

Hiby A.R.and A.J. Mullen (1980). Simultaeous determination of fluctuating age structure and mortality from field data. *Theor. Pop. Biol.* **18**: 192-203.

Hjort J. (1914). Fluctuations in the great fisheries of northern Europe. *Rapp. P.-v. Reun. Cons. int. Explor. Mer.* **20**:1-228.

de Hoog F.R. and M.F. Hutchinson (1987). An efficient method for calculating smoothing splines using orthogonal transformations. *Numer. Math.* **50**: 311-319.

Hutchinson M.F. (1986). Algorithm 642 A fast procedure for calculating minimum cross validation cubic smoothing splines. *ACM transactions on mathematical software* **12**: 150-153.

Hutchinson M.F. and F.R. de Hoog (1985). Smoothing noisy data with spline functions. *Numer. Math.* **47**: 99-106.

Hyman, J.M. (1983). Accurate monotonicity preserving cubic interpolation. *SIAM J. Scient. Stat. Comput.* **4**: 645-654.

Klein Breteler W.C.M., H.G. Fransz and S.R. Gonzalez (1982). Growth and development of four calanoidcopepod species under experimental and natural conditions. *Netherlands Journal of Sea Research* **16**:195-207.

Lancaster P. and K. Salkauskas (1986). Curve and Surface Fitting. Academic Press, London, UK.

Landry M.R. (1983). The development of marine calanoid copepods with comment on the isochronal rule. *Limnol. Oceanogr.* **24**:614-624.

Lynch M. (1982). How well does the Edmondson-Palohiemo model approximate instantaneous birth rates? *Ecology* **63**: 12-18.

Lynch M. (1983). Estimation of size-specific mortality rates in zooplankton populations by periodic sampling. *Limnol. Oceanogr.* **28**: 533-545.

McCauley, E., D. Hall, H. Caswell, A.M. de Roos, T. Hallam, S.A.L.M. Kooijman, M. Leibold, W.W. Murdoch, W. Neill, and R.M. Nisbet (1990). Population dynamics and modelling *in* Future Directions in Zooplankton Biology (A. Tessier and C. Goulden, Eds.), Princeton University Press, Princeton, USA.

McLaren I.A. (1978). Generation lengths of some temperate marine copepods: estimation, prediction and implications. *J. Fish. Res. Board Canada* **35**:1330-1342

Manly B.F.J. (1987). A multiple regression method for analysing stage frequency data. *Res. Pop. Ecol.* **29**: 119-127.

Meir A. and A. Sharma (1973). Spline functions and approximation theory. Birkhauser Verlag, Basel, Stuttgart.

Meyer S.L. (1975). Data Analysis for Scientists and Engineers. John Wiley and Sons, New York.

Mozorov, V.A. (1984). Methods for Solving Incorrectly Posed Problems. Springer Verlag, New York, USA.

Nisbet R.M. and Gurney W.S.C. (1982). Modelling Fluctuating Populations, John Wiley and Sons, Chichester and New York.

Nisbet, R.M., Gurney, W.S.C., Murdoch, W.W., McCauley, E. (1989). Structured population models: a tool for linking effects at individual and population level, *Biol. J. Linnean Soc.*, 37: 79-99 .

Palohiemo J.E. (1974). Calculation of instantaneous birth rate. *Limnol. Oceanogr.* 19: 692-694.

Paloheimo J.E. and W.D. Taylor (1987). Comments on life table parameters with reference to *Daphnia pulex. Theor. Pop. Biol.* 32: 289-302.

Parslow J., N.C. Sonntag and J.B.L. Matthews (1979). Technique of systems identification applied to estimating copepod population parameters. *J. Plankton Res.* 1: 137-151.

Poggio, T. and F. Girosi (1989). A theory of networks for approximation and learning. AI memo no 1140, Artificial Intelligence Laboratory, Massachusetts Institute of Technology, USA.

Press W.H., B.P. Flannery, S.A. Teukolsky and W.T. Vetterling (1986). Numerical Recipes. Cambridge University Press Cambridge, UK.

Ragozin D.L. (1983). Error bounds for derivative estimates based on spline smoothing of exact or noisy data. *J.Approximation Theory* 37: 335-355.

Reinsch C.H. (1967). Smoothing by spline functions. *Numer. Math.* 10: 177-183.

Rigler F.H. and J.M. Cooley (1974). The use of field data to derive population statistics of multivoltine copepods. *Limnol. Oceanogr.* 19: 636-655.

Romanovsky Y.E. and L.E. Polishchuck (1982). A theoretical approach to the calculation of secondary production at the population level. *Int. Revue ges. Hydrobiology* 67: 341-359

Saunders J.F. and Lewis W.M. (1987). A perspective on the use of cohort analysis to obtain demographic data for copepods. *Limnol. Oceanogr.* 32: 511-513.

Sazhina, L.I. (1968). On individual fecundity and duration of development of certain mass pelagic copepoda of the Black Sea. *Gidrobiologicheskii Zhurnal* 4: 69-72. [in Russian - cited by Corkett and McLaren (1978)].

Schneider S.M. and H. Ferris (1986). Estimation of stage-specific developmental times and survivorship from stage frequency data. *Res. Population Ecol.* 28: 267-280.

Seitz A. (1979). On the calculation of birth rates and death rates in fluctuating populations with continuous recruitment. *Oecologia* 41: 343-360.

Silverman, B.W. (1985). A fast and efficient cross-validation method for smoothing parameter choice in spline regression. *J. Am. Stat. Assoc.* **79**: 584-589.

Sonntag N.C. and J. Parslow (1981). Technique of systems identification applied to estimating copepod production. *J. Plankton Res.* **2**: 461-473.

van Straalen N.M. (1986). The 'inverse problem' in demographic analysis of stage structured populations. in The Dynamics of Physiologically Structured Populations (Lecture notes in Biomathematics 68: 393-408) Springer Verlag, Heidelberg, Germany.

Talenti, G. (Ed.) (1986). Inverse Problems, Springer Verlag Lecture Notes in Mathematics, Vol 1225.

Thompson B.M. (1982). Growth and development of *Pseudocalanus elongatus* and *Calanus sp.* in the laboratory. *J. Mar. Biol. Assoc. U.K.* **62**: 359-372

Tikhonov, A.M. (1963). On the solution of ill-posed problems and the method of regularisation. *Soviet Math. Dokl.* **4**: 1035-1038 (English translation).

Tukey J.W. (1977). Exploratory Data Analysis. Addison Wesley, Reading, Mass. USA.

Twombly, S. (1983). Paterns of abundance and population dynamics of zooplankton in tropical Lake Malawi. PhD thesis, Yale University.

Utreras F.I. (1985). Smoothing noisy data under monotonicity constraints existence, characterization and convergence rates. *Numer. Math.* **47**: 611-625.

Wahba G. (1981). Numerical experiments with the thin plate histospline. *Communications in Statistics Series A* **10**: 2475-2514.

Wahba G. (1983). Bayesian "Confidence Intervals" for the cross-validated smoothing spline. *J. Roy. Stat. Soc. Series B* **45**: 133-150.

Wahba G. and Wold S. (1975). A completely automatic french curve: fitting spline functions by cross validation. *Communications in Statistics* **4**: 1-17.

Wood, S.N. (1989). Estimation of Mortality Rtaes in Stage Structured Zooplankton Populations. PhD Thesis, University of Strathclyde, Glasgow, UK.

Wood, S.N, S.P. Blythe, W.S.C. Gurney, and R.M. Nisbet (1989). Instability in mortality estimation schemes related to stage structure population models. *IMA J. Math. Appl. Med. Biol.* **6**: 47-68.

Yan Z. (1987). Piecewise cubic curve fitting algorithm. *Mathematics of Computation.* 49: 203-213.

P. Dallos, C. D. Geisler, J. W. Matthews,
M. A. Ruggero, C. R. Steele (Eds.)

The Mechanics and Biophysics of Hearing

Proceedings of a Conference
Held at the University of Wisconsin, Madison,
WI, June 25–29, 1990

1991. VII, 418 pp. (Lecture Notes in Biomathematics, Vol. 87) Softcover DM 80,–
ISBN 3-540-97473-3

Proceedings of a workshop on the physics and
biophysics of hearing that brought together
experimenters and modelers working on all
aspects of audition.
Topics covered include: cochlear mechanical
measurements, cochlear models, mechanicals
and biophysics of hair cells, efferent control, and
ultrastructure.

R. H. Bradbury, National Resource Information
Centre Canberra, A. C. T. (Ed.)

Acanthaster and the Coral Reef: A Theoretical Perspective

Proceedings of a Workshop
Held at the Australian Institute of Marine
Science, Townsville, August 6–7, 1988

1990. VI, 338 pp. (Lecture Notes in Biomathematics, Vol. 88) Softcover DM 61,–
ISBN 3-540-53501-2

The mathematical analysis of an outbreaking
species, the crown-of-thorns starfish found on
Indo-Pacific coral reefs was the central topic that
brought together mathematicians, ecologists and
oceanographers in an attempt to create a new
paradigm for understanding this complex pheno-
menon. A wide variety of mathematical
approaches was offered in the workshop, from
traditional qualitative stability analysis to work
on Finsler spaces, grammars and adaptive
systems. Together they point to a new understanding of the dynamics of the outbreaks and of
the stability of the coral reef ecosystems in which
they occur.

W. Alt, University of Bonn; G. Hoffmann,
University of Würzburg (Eds.)

Biological Motion

Proceedings of a Workshop
Held in Königswinter, Germany,
March 16–19, 1989

1990. X, 604 pp. (Lecture Notes in Biomathematics, Vol. 89) Softcover DM 128,–
ISBN 3-540-53520-9

The diverse aspects of Biological Motion were
the central topic that attracted the participants of
this workshop on Modeling, Analysis and Simulation. In various working groups they discussed
movements paths and searching behavior, kine-
sis and/or taxis, cellular motion and shape
changing, microtubuli and cilia motion, neuro-
motoric aspects of animal movement, collective
behavior and swarming. The resulting contributions to the Proceedings, in part strongly
influenced by these extensive and fruitful discus-
sions, were rearranged and complemented by
section overviews as well as "boxes" describing
the basic mathematical methods. This makes the
book a valuable reference for biomathematicians
as well as for theoretically interested biologists,
when trying to quantify motions of cells, organisms or parts of these.

Springer-Verlag
Berlin
Heidelberg
New York
London
Paris
Tokyo
Hong Kong
Barcelona
Budapest

Lecture Notes in Biomathematics

For information about Vols. 1–47
please contact your bookseller or Springer-Verlag

General Remarks

Lecture Notes are printed by photo-offset from the master-copy delivered in camera-ready form by the authors of monographs, resp. editors of proceedings volumes. For this purpose Springer-Verlag provides technical instructions for the preparation of manuscripts. Volume editors are requested to distribute these to all contributing authors of proceedings volumes. Some homogeneity in the presentation of the contributions in a multi-author volume is desirable.

Careful preparation of manuscripts will help keep production time short and ensure a satisfactory appearance of the finished book. The actual production of a Lecture Notes volume normally takes approximately 8 weeks.

For monograph manuscripts typed or typeset according to our instructions, Springer-Verlag can, if necessary, contribute towards the preparation costs at a fixed rate.

Authors of monographs receive 50 free copies of their book. Editors of proceedings volumes similarly receive 50 copies of the book and are responsible for redistributing these to authors etc. at their discretion. No reprints of individual contributions can be supplied. No royalty is paid on Lecture Notes volumes.

me authors and editors are entitled to purchase further copies of their book for their personal use at a discount of 33.3 %, other Springer mathematics books at a discount of 20 % directly from Springer-Verlag. Authors contributing to proceedings volumes may purchase the volume in which their article appears at a discount of 20 %.

Commitment to publish is made by letter of intent rather than by signing a formal contract. Springer-Verlag secures the copyright for each volume.

Addresses:

Professor Simon A. Levin, Cornell University
Section of Ecology and Systematics
345 Corson Hall, Ithaca
New York 14853-0239, USA

Springer-Verlag, Mathematics Editorial
Tiergartenstr. 17
W-6900 Heidelberg
Federal Republic of Germany
Tel.: *49 (6221) 487-410